The Investigation of
FIRES

CHARLES L. ROBLEE
Gateway Technical Institute, Racine

ALLEN J. McKECHNIE
Property Loss Research Bureau, Chicago

PRENTICE-HALL, INC.
Englewood Cliffs, New Jersey 07632

Library of Congress Cataloging in Publication Data

ROBLEE, CHARLES L (date)
    The investigation of fires.

    Bibliography: p.
    Includes index.
    1. Fire investigation. I. McKechnie, Allen J.,
joint author. II. Title.
TH9180.R62   1981   363.2'5   80-17224
ISBN 0-13-503169-9

Printed in the United States of America
10 9 8 7 6 5 4 3 2 1

Editorial/production supervision and interior design by
Maureen Wilson
Cover design by Wanda Lubelska
Manufacturing buyer: Gordon Osbourne

PRENTICE-HALL INTERNATIONAL, INC., *London*
PRENTICE-HALL OF AUSTRALIA PTY. LIMITED, *Sydney*
PRENTICE-HALL OF CANADA, LTD., *Toronto*
PRENTICE-HALL OF INDIA PRIVATE LIMITED, *New Delhi*
PRENTICE-HALL OF JAPAN, INC., *Tokyo*
PRENTICE-HALL OF SOUTHEAST ASIA PTE. LTD., *Singapore*
WHITEHALL BOOKS LIMITED, *Wellington, New Zealand*

To the memory of
those firefighters and fire investigators
who have lost their lives
in incendiary fires

Contents

v

CARE AND HANDLING OF PHYSICAL EVIDENCE   119

MOTIVES FOR FIRESETTING   129

INTERVIEWS AND INTERROGATIONS   143

RECORDING THE INVESTIGATION   156

13

A

B

Foreword

Our nation is finally recognizing the fact that the crime of arson has reached epidemic proportions. This fact is evidenced by the reclassification of arson as a Class I offense by the Federal Bureau of Investigation. Fire and law enforcement agencies must do everything in their power to combat this heinous crime.

Mr. Roblee and Mr. McKechnie have combined their talents to coauthor a no-nonsense book covering many aspects of fire investigation. This book will be a valuable asset to the novice as well as the experienced investigator.

Chuck Roblee has over 20 years' teaching experience in fire investigation and 33 years of practical experience in fire and accident investigation.

Al McKechnie is a Special Agent with the Property Loss Research Bureau and was formerly a Detective Sergeant of the DuPage County, Illinois, Sheriff's Department.

RONALD W. CHIAPETE, *Chief*
Racine, Wisconsin, Fire Department

Preface

This is a basic book on fire investigation, *not* a text on hazardous materials or on firefighting tactics. However, several chapters have been written with the needs of the fire-science- or fire-service-oriented person in mind. The chapter on the chemistry of fire and fire behavior will be a review for most of them, and it will introduce the subject to the law-enforcement-oriented reader. Many of the latter are not familiar with this subject, the knowledge of which is vital to the investigation of fires.

Recently, the U.S. Supreme Court handed down a landmark decision on search and seizure in the investigation of fires and arson. In *Michigan* v. *Tyler et al.*, the Court applied the three types of searches set forth in the *Camara* and *See* cases. These decisions and several others, together with what constitutes the *corpus delicti* of the crime of arson, are discussed in the chapter on the law of arson.

Following these basic chapters, the text discusses the observations that should be made by firefighters and police officers prior to and after their arrival at the fire scene. Until the fire investigator arrives there, the firefighters and police officers are his eyes and ears; what they see and hear can be of tremendous value.

Chapter 5 describes the procedures that may be used to search a fire scene. It discusses the use of burn patterns and char depths, the result of combustible fuels' heating or partially burning, as burn indicators. The text describes how these indicators may be used in interpretation, as that is one of the primary means of determining the causes of fires.

The courts recognize three basic causes of fire: natural, accidental, and incendiary. Fire causes are discussed within this frame of reference in Chapter 6.

Most texts on fire and arson investigation do not treat the subject of explosions to any great extent. However, in most explosions, it is the fire investigator who makes the initial investigation. In many jurisdictions, unless the explosion was of a non-accidental cause, such as a bomb, the entire investigation is the responsibility of this official. The chapter on explosions discusses the types, sources, causes, and characteristics of explosions, and the meaning of some of the patterns and results found after an explosion has occurred.

Fires involving deaths present additional problems for the investigator. These are enumerated for the reader in the chapter on fatal fires. The importance of an autopsy and post mortem investigation in these cases cannot be overemphasized.

Probably the most interesting topic in connection with the investigation of fires is, "Why was the fire set?" The motives for setting fires are numerous, and investigation requires an understanding of the perpetrator's intent. The chapter on motives examines the many and varied reasons for firesetting.

Chapters 11 and 12 discuss the techniques of interviewing people and recording information, the "meat and potatoes" phases of the fire investigator's work. The final chapter takes up the matter of arson techniques, to convey in some detail the specifics that the investigator must look for in a suspicious fire.

This book has been written in a straightforward manner with a reading level ideally suited for readers of diverse backgrounds and education. The material included is fundamental to the making of a fire investigation.

Acknowledgments

The authors wish to express their sincere appreciation to those colleagues and others who assisted in the preparation of this book.

Special thanks should be given to Robert Stedman, who reviewed the original manuscript.

Our thanks also to Doris Myers for her analysis of the reading levels of each chapter; to James Markusen, Michael Cooper, and William Jones for their assistance in illustrating the book; and to William G. Hansen and Ed Roblee, who prepared the line drawings.

We are grateful to Laurence R. Simson, Jr., M.D., Forensic Pathologist, for his review and suggestions for the chapter, "The Fatal Fire," to James McCaskey for his suggestions on interviewing and interrogations, to John C. Burn for his assistance on recording the investigation, and to Thomas A. Kuehn for his comments on the chapter on explosions.

To Dick Roblee, our thanks for his many suggestions and comments during the preparation of the manuscript.

Our Fire Science students offered encouragement throughout the time we spent in writing the book.

And finally, to our wives, Millie and Barbara, deepest gratitude for their patience, understanding, and encouragement.

CHARLES L. ROBLEE
ALLEN J. McKECHNIE

Why Fires Should Be Investigated

Incendiarism has plagued the human race ever since it discovered fire. These days, firesetting and arson seem to be on the increase. Some "experts" deny there is such an increase, but those in the fire service and particularly those who investigate fires take issue with these "experts." Others speculate as to whether it is the number of incendiary fires or the recognition of more fires as incendiary that has grown. In any event, no one will deny that the problem exists. In fact, after many years of petitioning by various organizations, Congress finally mandated in 1978 that the Federal Bureau of Investigation add arson to its list of major crimes.

The detection of arson cases can result only from the thorough investigation of all fires. Basic fire investigation is the prelude to an arson investigation.

The Law Enforcement Assistance Administration's report entitled *Arson and Arson Investigation: Survey and Assessment* reported on a survey of current needs in arson investigation. Of first priority was "increased training for arson investigators and for judges and prosecutors in the technicalities of arson cases."[1] The book you are reading, which represents the authors' combined experience of 40 years in investigating fires and 26 years of teaching courses on the investigation of fires, will, we hope, assist in fulfilling that priority.

Several years ago, speaking at an International Arson Investigators Seminar at Purdue University, Captain Sam S. Cobb, Jr., past president of the International Association of Arson Investigators, discussed the responsibilities of the fire investigator. What he said at that time has not changed over the past years.

---

[1] John F. Boudreau, Quon Y. Kwan, William E. Faragher, and Genevieve C. Denault, *Arson and Arson Investigation: Survey and Assessment* (Washington, D.C.: U.S. Government Printing Office, 1977), p. xv. Preparation of this document was supported by a grant awarded by the National Institute of Law Enforcement and Criminal Justice to the Aerospace Corporation.

Mr. Investigator, . . . your responsibility is an important facet of the arson case. You are the "fact finder" and "seeker of the truth." Once you have been alerted by the firefighter that the fire is of suspicious or incendiary origin, your duty is to establish the facts, to find, interpret, correlate and evaluate the evidence in the case.[2]

As "seeker of the truth," the fire investigator must always seek the true cause of the fire. This is his or her first responsibility. Of the three basic causes of fire recognized by the courts, it is the fire investigator's obligation to determine which applies in any given case. (These basic causes are enumerated in Chapter 3.) To accomplish this, the fire investigator must proceed in a manner similar to that of any other investigator: He must observe and record the facts found at the fire scene. Some of the methods that may be used are discussed in Chapter 12, "Recording the Investigation."

Arson has become a very sophisticated operation. "Arson is one of the most difficult crimes to prevent, detect, investigate and prosecute successfully," says the "First Report of the Mayor's Task Force on Arson" of Seattle, Washington.

There are many ways to attack the arson problem through government agencies. Some of these methods have proved more successful than others. For example, the concept of a special arson squad made up of members of both fire and law-enforcement personnel is not new; it has been used for many years in a number of communities and municipalities. Detroit, Michigan, had good results in the 1950s and 1960s from such an arson squad, which was under the supervision of Detective Inspector Glenn Bennett, former president of the International Association of Arson Investigators.

The current fire and arson investigational problem is of epidemic proportions. It has become so large and so complex that a more sophisticated approach is required. Today's investigative unit is only one segment of the larger organization developed to fight the crime of arson. Basically, this group is an expansion of the combined arson squad. It has become known as the "arson task force."

The arson-task-force concept has been developed and refined over the past few years. It is usually composed of representatives from (1) the insurance industry; (2) those responsible for prosecution of crimes; (3) the local government, both the executive and legislative branches; (4) the business community (Chamber of Commerce); (5) the combined arson squad; and (6) and any other law-enforcement agency. These representatives work closely together, planning, developing, and directing an all-out attack on arson. Some of the methods used are 24-hour hot lines and monetary rewards for information. All participating segments of the task force provide the investigational unit with internally

[2]Samuel S. Cobb, Jr., "Responsibility of the Firefighter and Investigator in Arson Cases," mimeographed paper delivered at the 19th Annual International Arson Investigators Seminar, West Lafayette, Ind., Public Safety Institute, Purdue University, April 22, 1963, p. 5.

developed intelligence. The results in some communities have been spectacular, with arson dollar losses dropping by 30 percent or more.

But the arson task force alone is not the entire answer to the arson problem. It has too many facets. In addition to the efforts of the task force, others must become involved. For instance, the cooperation of many individuals, private interests, federal agencies, and local governments is necessary to retard and prevent the deterioration of neighborhoods that has led to systematic destruction of buildings by arson. Some steps have been taken in various parts of the country to obtain this cooperation, mostly through the efforts of an arson task force. Cooperation on the federal level has resulted in the designation of the Law Enforcement Assistance Administration (LEAA) as the agency to support local and state governments, as well as private agencies, through the transfer of funds. The use of materials developed by the U.S. Fire Administration (USFA) and other federal agencies is encouraged. In the final analysis, the arson task force, in order to be successful, must involve the entire community in its effort to prevent and suppress the crime of arson.

The investigation of all fires is basic to the discovery of an incendiary fire. A fire loss of $100,000 may make the front page of the local paper and draw one or two fire investigators to the scene. On the other hand, a bank robbery involving the same amount of money will make headlines across the nation and mobilize a great many law-enforcement personnel and agencies. But which event causes the most personal suffering and anguish? In which is the loss of life most likely to occur?

Fire losses in the United States are in excess of $6 billion annually. Over 10,000 fire-caused fatalities and more than 13,000 nonfatal injuries occur each year. To prevent this loss of life and property and this suffering, fires must be prevented. Only through the proper investigation of each fire can the true cause be determined. And only after the causes are known can steps be taken to prevent similar fires from occurring. Trained fire investigators in all communities and municipalities are a must.

The first step toward a successful fire investigation is the thorough understanding of the chemistry of fire and fire behavior, the subject of the next chapter.

The Chemistry of Fire and Fire Behavior

To be a successful fire investigator, one must understand the many properties and characteristics of fire, as well as of people and investigational techniques. The fire investigator must thoroughly comprehend the chemistry of fire and fire behavior—the properties of fire, how it burns, and how it behaves. If he or she does not, his or her success in determining the origin and cause of fires will be only incidental or accidental. Not all fires burn in the same manner, since each has its own environment. Many times, it is the environment that determines how a fire will behave.

The Chemistry of Fire

Fire, like all other natural phenomena, follows the laws of nature. There are three elements of fire: fuel, heat, and oxygen. These form what is commonly known as the *fire triangle*. The term *fire tetrahedron* has been used to describe the production of free radicals during the combustion process. This phase of fire behavior and the chemistry of fire is very important to the student of fire suppression. Without an understanding of its principles, the suppression characteristics of some of the most important extinguishing agents cannot be understood.

Fuel is probably the most important part of the fire triangle to the fire investigator, for fuel is what burns. Therefore, the nature and properties of fuel become essential information.

Fuels come in three forms: solid, liquid, and gaseous. All are affected by heat, but each has its own characteristics. Solid fuels have definite volume and shape. Liquid fuels have definite volume, but no shape; they assume the shape of whatever contains them. Gases have neither definite shape nor volume; they expand and contract and assume the shape and volume of any vessel in which they are confined. Fuels can exist in any of these forms.

Once the fuel becomes ignited, the characteristics of the fire that follows depend upon the chemical makeup of the fuel. The process whereby fire consumes most solids is called *pyrolysis*. Pyrolysis is the result of heat's being applied to a solid fuel. As the fuel becomes heated, the moisture in it begins to produce water vapors. Shortly thereafter, the decomposition begins to produce combustible vapors. It is these vapors that burn. Most solid organic compounds, such as wood, plastic, or coal, do not burn; they actually pyrolyze. It is the combustible products of their pyrolytic decomposition that burn, and as these volatile gases burn, they rise.

By understanding pyrolysis, the investigator can understand how fire and heat are transferred, and why fuel above the fire becomes ignited while that below does not burn as easily. Without the vaporous materials produced by pyrolysis from the solid fuel—the materials that burn—the fire would be nonexistent.

PYROLYSIS OF WOOD

One of the major components of wood is cellulose. It is this component that is pyrolyzed. As the temperature of the wood rises, it begins to decompose chemically, with water vapors being the first volatile produced. As the process continues, carbon monoxide, carbon dioxide, hydrogen, and more water and other vapors are distilled. Finally, all that remains is carbon in the form of charcoal. The soft woods, particularly the pines, contain large quantities of resins in their cellular pockets. When exposed to heat, these resins tend to vaporize rapidly. And because soft woods have a larger cellular structure than the hardwoods, the combustible vapors are distilled at a lower temperature than those found in hardwoods. This accounts for the difference in the ignition temperatures of the various woods. Figure 2-1 illustrates the pyrolytic process in wood.

Pyrolysis, or the thermal decomposition of wood, may also occur in the absence of oxygen. It is a relatively slow process. As the heat produces the products of pyrolysis, combustible and noncombustible gases and vapors are distilled. The end product is almost solid charcoal. The competent investigator will recognize a flat, baked appearance, with many fire craze lines, as an indication of a very long, low heat. This flat, baked appearance will be found throughout the wood pyrolyzed under these conditions. Figure 2-2 shows this type of charring.

This particular type of pyrolysis can occur at temperatures below the ignition temperature of the wood. Wood, in close proximity to steam lines, has been found to have become charred over a period of years as the heat has gradually produced pyrolytic decomposition. The heating does not have to be continuous to produce this effect; it is the total time of exposure to the heat that does it.

In contrast to this flat, baked appearance, large, rolling blisters indicate the rapid, intense heat of a fast-burning fire. Figure 2-3 is illustrative of this type of charring or fire pattern.

The shape of solid fuels can affect their ability to ignite. Solids burn only on the surface exposed to air. The greater the surface area, the more likely igni-

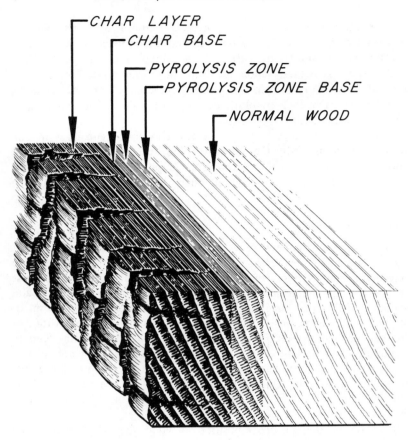

FIGURE 2-1. The process and steps of the pyrolytic decomposition of wood. [Source: E. L. Schaffer, *Charring Rate of Selected Woods—Transverse to Grain*, U.S. Forest Service Research Paper FPL 69 (Madison, Wis.: U.S. Forest Products Laboratory, 1967), p. 2.]

tion and combustion will take place. It is easier to ignite a piece of kindling than a two-by-four; it is easier to ignite wood shavings than kindling. The dusts of combustible solids can explode when they are suspended in air and ignited.

Pyrolysis of Liquid Fuels

Theoretically, liquid fuels are also subject to pyrolysis. Flammable and combustible liquids such as gasoline, fuel oil, and turpentine are volatile. They may vaporize rapidly when heated, and the vapors burn. Most of the time, these liquid fuels have a boiling point that is lower than the temperature at which they would pyrolyze. Therefore, the boiling process produces the vapors and gases before the pyrolytic temperature is reached. In most cases, the liquid fuels of interest to the fire investigator fall within this category. They are not pyrolyzed in a fire, but are distilled into it by the boiling process.

FIGURE 2-2. Wood completely pyrolyzed from long exposure to relatively low heat. Note flat, baked appearance, with fire craze lines.

COMBUSTION AND HEAT

Oxidation is the chemical union of a substance with oxygen or an oxidizing agent. Combustion or fire is rapid oxidation. Heat, the principal product of fire and an element of it, is a form of energy. Heat energy is the result of the movement or vibration of the molecules of a substance. The greater the vibration, the more heat produced. The intensity of the vibration is measured by the temperature or degrees of heat. As the molecules of the substance vibrate, they bump or come in contact with other molecules, causing them to move. As this process continues, the heat spreads farther and farther within the substance being heated.

Combustion cannot occur below the ignition temperature of the fuel. Ignition temperature and flash point are often confused. *Flash point* is the temperature at which a flammable liquid produces sufficient vapors to be ignited by a flame, spark, or glowing fire. It should be understood that the vapors do not have to continue to burn at the flash point, but they will flash. *Ignition temperature* is the minimum temperature required to initiate a self-sustained combustion that is independent of any heating element. Ignition temperature applies to

FIGURE 2-3. Large, rolling blisters, indicating fast-burning fire with intense heat.

all types of fuels—solid, liquid, or gaseous. Flash point applies only to flammable liquids and a very few solids. Two solids that do have a flash point are naphthalene and camphor.

As an example, assume that a beaker of flammable liquid is placed in an oven, and that heating elements are independent of that part of the oven where the beaker rests. When the temperature of the flammable liquid reaches its ignition point, the flammable liquid will automatically ignite, but no spark, flame, or glowing fire is present.

The exact ignition temperature of an ordinary solid fuel may vary as the result of its moisture content. Wet or damp duel is difficult to ignite.

Fire Behavior as Applied to Fire Investigation

HEAT TRANSFER

Fire communicates and heat is transferred by three methods: (1) conduction, (2) convection, and (3) radiation.

Conduction · Heat travels by conduction through solids or between solids in contact with each other. The concept of the vibration of molecules makes it easy to visualize how heat will flow in a solid. The rate of conduction depends upon the temperature difference and density of the solid.

Convection · Heat travels by convection by moving from one molecule to another, but the particles or molecules are themselves in motion. Gases, in particular, being heated tend to circulate and spread the heat. Convection currents may produce the "fireball," a phenomenon usually generated by a very intense fire. The combustibles produced by the fire rise in the convection current. They burn only around the periphery of the combustible "fireball," since the mixture within is too rich to burn.

Convection currents also carry combustible and heated pyrolytic products upward in a structure to the highest point available. The gases then mushroom out across the ceiling or bottom side of the roof until their horizontal movement is blocked. At this point, they begin to bank down along the sides or walls of the structure. When the ignition temperature of these gases is reached, they flash and ignite into what is commonly called the *flashover*.

In structural fires, heat normally rises within the buildup of the convection currents or thermal column. The cooler air begins to settle to the lower levels. Temperatures at the ceiling, when the flashover occurs, will vary with the combustibility of the contents of that particular area.

In a fire in the ordinary dwelling type of occupancy, the temperatures at the ceiling level will range from 1,000°F to 1,600°F. In the same room, the temperature at the floor level will be from 150°F to 250°F. The differences between these temperatures will depend upon the amount of smoke or steam present. Radiant heat is not likely to penetrate smoke and steam, but in a room or area with a minimum amount of smoke, the radiant heat can cause charring and burning of the combustible materials at the floor level. Char and burn patterns developed under these conditions could easily mislead the uninformed investigator.

Radiation · Heat energy is radiated from a hot body to a cold body. Light is a visible form of radiation.

Radiant heat has been responsible for the communication of fires between buildings almost 100 feet apart. The radiation is absorbed by the combustible surfaces of the exposed structure. As the heat accumulates, pyrolytic decomposition begins. When the ignition temperature of the products produced by the pyrolysis is reached, the material will flash into flames.

Heat may travel by conduction and radiation in any direction; convection currents always tend to rise. Almost everyone has seen, at one time or another, the smoke from a campfire, rubbish fire, or smoke stack rise high in the air. This is a visible example of the convection current or thermal column of the fire.

As convection currents rise within a structure, they are blocked by the ceiling. This causes them to move horizontally until they are stopped by a wall or find another vertical opening. If there is no vertical opening, the wall or another obstruction stops the horizontal movement, so they build up and begin to bank downward. Here is where the structural environment of the room or area becomes important. Built-in cabinets, soffits, arches in hallways or between rooms, or dropped ceilings can direct the heat and products of combustion.

The paths and burn patterns established by these construction features may cause confusion to the investigator. The feature that produced the burn pattern may have been destroyed by the fire, by overhaul practices, or by suppression efforts of the fire department.

THERMAL BALANCE AND THERMAL IMBALANCE

Thermal balance and thermal imbalance are two other types of fire behavior the investigator must be familiar with. *Thermal balance* is the normal pattern or movement of fire, smoke, and fire gases within a structure—the normal or natural condition created by the fire. As products of combustion rise in a structure, or flow out of an opening, an equal volume of air replaces then. If, during extinguishment, water is distributed in such a manner as to upset this balance, a condition known as *thermal imbalance* may occur. Hot spots may develop in the center of the area. Turbulent circulation of steam and smoke may replace the normal flow of the products of combustion.

It was reported that in test fires in Des Moines, Iowa, in November 1959, a line of holes was burned in the floor, from three to six feet from the nearest fuel cribs. They were determined to have been burned through the floor from the top down. It was concluded that heat pockets had been trapped in such a manner that fuel decomposition in them was at a rate in excess of that ordinarily found in the free-burning phase.[1] This type of fire pattern, developing from thermal imbalance, could easily mislead a fire investigator into believing that a flammable-liquid pattern existed on the floor.

THE PHASES OF FIRE

A fire is usually considered to consist of three phases: the initial or incipient phase, the free-burning phase, and the smoldering phase.

Incipient phase · In the incipient phase, the base area of the fire has a temperature of 400°F to 800°F. The room has a normal temperature, and as the heat from the fire rises, the ambient temperature cools it until the ceiling temperature is about 200°F. The products of pyrolysis are mostly water vapors and carbon dioxide in this phase. Minute quantities of carbon monoxide and sulphur dioxide may also be present. This is the beginning of a fire.

---

[1] "Firemanship Training," *Test Fire, Des Moines, Iowa, November 14, 1959* (Ames, Ia.: Engineering Extension, Iowa State University, n.d.) pp. 13–14.

Free-burning phase · As the fire continues to burn and build up heat, the pyrolytic process accelerates. The thermal column begins to develop and the heat rises. Temperatures in the base area of the fire may reach 800°-1,000°F, and at the ceiling, 1,200°-1,600°F. If the fire is in the vicinity of a wall, the rising pattern of the thermal column can be apparent on the wall. As the heat builds up, the pyrolytic decomposition moves upward along the wall. As it rises, it fans out. The wall and any fuel in its immediate vicinity begin to char and pyrolyze.

Here, we perceive the V or inverted-cone pattern created by the fire. Figure 2-4 is a picture of this type of pattern. The bottom of the inverted cone indicates the point of origin of this combustion. In the picture, the pattern begins near the floor and spreads up the wall. It has consumed the shelving and all the combustibles available.

It is during this free-burning phase that the maximum heat and destructive capabilities of the fire develop. The thermal column carries the pyrolytic products upward until they are stopped by the ceiling. Here, the heat concentrates directly above the fire, gradually heating and pyrolyzing the combustibles in that area. Some of the heat and burning products of combustion begin to mushroom out from this point. When the heat has brought the combustible portions of the ceiling to their ignition temperature, the flashover will occur.

FIGURE 2-4. Inverted-cone or V pattern on wall illustrates fire rising from point of origin.

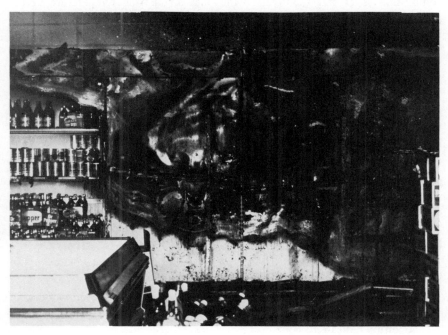

During the free-burning phase, there is little accumulation of the products of combustion, because most of them are being consumed by the flames. The fire continues to burn and spreads from within the room of origin to other rooms or areas, igniting fuel sources farther and farther away. But the fire in the original area continues to burn, destroying more and more of the structure and contents of it. This area is usually indicated by deep charring or nearly total destruction and is the area searched for by the fire investigator.

The recognition of the typical fire patterns developed during this phase can be of utmost importance to the investigator, whose understanding of these patterns helps him to determine the area of origin of the fire. And ultimately they lead him to the point or points or origin. It is the ability to locate the pattern or patterns of burning and correctly interpret them that enables the experienced investigator to do his or her job.

Smoldering phase · Should the air and oxygen become exhausted in the area of the fire, other changes take place in the fire behavior. When the oxygen content drops to below 13 percent, the fire can no longer support flames. As the flame production ceases, the fire begins to smolder and glow. With no flames, the thermal column ceases and the heat begins to develop into layers. The temperatures in the room drop. The ceiling temperature will be about 1,000°-1,300°F. The products of combustion not being consumed by the flames continue to build up in volume. The most likely product of this incomplete combustion is carbon monoxide, which has an ignition temperature of approximately 1,125°F. In addition, the pressures within the room build up. The combined gas law (a combination of Charles' Law and Boyle's Law) becomes effective. For every rise of approximately 460°F within a confined area, the volume of gases is doubled. And when the volume of gases in a confined area is doubled, the atmospheric pressure within the area will double.

At this point, the fire area contains large quantities of superheated fuel under pressure, but little oxygen. All that is needed for the area to explode is the introduction of the necessary air or oxygen. This type of explosive condition can result in a backdraft.

Fire is usually considered to be in one of two forms. In the first and second phases it is a flaming fire; in the third phase, it is the glowing fire of the charcoal, which is the hot carbon remaining after pyrolysis. These glowing charcoal embers depend upon any oxygen moving through the smoldering fire. As this air is minimal, so is its oxygen content; as a result, these glowing coals produce large quantities of carbon monoxide.

QUESTIONS

1. As a fire investigator, you observe a large quantity of charred wood with large, rolling blisters in the area of the origin of the fire. What does this condition indicate to you?

2. The woodwork and all wooden furniture in the upper half of the room of origin has a flat, baked appearance, with many fire craze lines. What can you deduce from this condition?

3. Why is the structural environment important to the development of burn patterns?

4. Describe the three methods by which heat may be transferred and by which fire may communicate.

5. What is meant by thermal balance? thermal imbalance?

6. What is the process by which carbonaceous materials, such as wood, are pyrolyzed?

7. In most cases, the liquid fuels of interest to the fire investigator do not pyrolyze. Why?

8. Discuss the inherent dangers connected with the smoldering phase of fire.

The Law of Arson

Law may be classified in several ways. One such classification is into written and unwritten law. Written law consists of those rules of conduct that have been formally enacted by a legislative (lawmaking) body. Examples of this type of law are constitutions and statutes.

Unwritten law consists of those rules of conduct that are binding but have not been enacted by a legislative body. Examples are the common law embodied in decisions of the courts. In this case, these rules may have to be derived from a series of decisions by the process of induction. Custom, if recognized by the courts as binding, may also be a form of unwritten (common) law.

The classification of law into written and unwritten also suggests a classification of the sources of law. Formally adopted enactments, such as statutes and ordinances, are important sources of law in modern times. Unwritten rules embodied mainly in court decisions are an even more important source of law.

One usually thinks of common law as "judge-" or "bench-made law." This type of common law has evolved from legal practices in medieval England. All the states of the United States, with the exception of Louisiana, derive their common law from England. Louisiana's law comes from the Code Napoleon (1804).

Common Law and Precedent

The principles of precedent and common law are inseparable. The Latin phrase *stare decisis*, meaning "Let past decisions stand," is the basis of precedent. Indeed, precedent is the keystone of the common law. This fact requires the investigator to look carefully at court decisions and court interpretations of the written law.

The practice of judges of adhering to decisions of the past gives a degree of stability and predictability to the administration of justice. In cases that break new ground, the judge has the obligation to weigh carefully the advantage of holding to the precedents of legal history against change.

TYPES OF PRECEDENTS

Precedents are usually said to be either "compelling" or "persuasive." Compelling precedents are those set or established within the judicial jurisdiction of which the court considering the precedent is a part. As an example, a decision or precedent set in a U.S. district court will be considered to be binding only upon the U.S. district courts of that particular district. A decision by the Seventh U.S. Court of Appeals would be considered a compelling precedent throughout the Seventh Judicial District (Illinois, Indiana, and Wisconsin).

By the same token, a decision handed down in the Seventh U.S. Court of Appeals *will not* be compelling in any other U.S. judicial district. Rather, it would be "persuasive." A district court in another judicial district might cite it as a precedent if it wished. However, there could be no compelling precedent within the district that was overriding.

A precedent established by the U.S. Supreme Court becomes compelling in *all* federal courts and in *all* state courts where the particular point is in question. Occasionally, the Supreme Court will reverse a previous decision. This establishes a new precedent or series of precedents. Probably the most famous of these changes occured in *Brown* v. *Board of Education of Topeka*, and in the *Escobedo* and *Miranda* cases. In regard to fire investigation activities, the U.S. Supreme Court handed down a decision in *Michigan* v. *Tyler*[1] in 1978 (see Appendix B). This case applied the principles set forth in *Camara* v. *Municipal Court of the City and County of San Francisco*[2] and *See* v. *City of Seattle*.[3] *Camara* had established the guidelines for regulatory, administrative, and criminal searches and inspections. These cases will be discussed later in this chapter.

Legal Definitions of Arson

In general, the common-law definition of arson was traditionally the willful and malicious burning of the house of another, including all outhouses or outbuildings adjoining thereto. The emphasis was on another's habitation, and his life and safety at the place where he resided. Then, many legal issues began to arise. Was a school a dwelling? a jail? a church? the common-law courts began to view the crime of arson as being against the habitation *or possessions* of another.

Gradually, laws were enacted to plug the loopholes of the common-law definition of arson. The first laws brought all buildings or structures into the scope of arson, provided they had human occupancy of any kind on a regular

[1] 436 U.S. 499, 56 L. Ed. 2d 486 (1978).
[2] 387 U.S. 523, 87 S.Ct. 1727 (1967).
[3] 387 U.S. 541, 87 S.Ct. 1737 (1967).

basis. Later, the occupancy requirements were dropped. Today, *arson* is a term applied to the willful and intentional burning of all types of structures, vehicles, forests, fields, and so on.

Most states have laws that define arson as the willful and intentional damage or destruction of property. Arson is usually classified in two or more degrees. Its seriousness and the penalties depend upon several factors. The type of property involved, whether it is occupied (usually a more serious offense), whether the arson is commited at night (more serious) or during the daytime, and the danger to human life are all taken into account. In addition, there are usually laws dealing with attempted arson and arson with intent to defraud insurance companies.

The *Corpus Delicti* of Arson

Although laws vary from state to state, the primary problems faced by the investigator are similar. The most important task is to *determine the true cause of the fire*. If necessary, the *corpus delicti* of the crime of arson must be established. *Corpus delicti* means "body of the crime," or the elements of the crime.

Fires have three general causes:

1. An act of God or providential origin
2. An accidental origin
3. An incendiary origin

The investigator must establish in court, beyond any reasonable doubt, the *corpus delicti* of the crime of arson. Curtis states:

> The mere burning of a building does not constitute the *corpus delicti* of the crime of arson. There is no presumption that a burning building has been intentionally set on fire; on the contrary, the presumption of innocence, which is accorded to an accused, carries with it a presumption that the fire is of accidental or providential origin.[4]

Therefore, it is necessary for the investigator to eliminate all possible natural (act of God) and accidental causes. For if the defense can introduce *any possible* accidental or natural cause, a directed verdict of not guilty can be forthcoming.

Curtis continues his discussion of this point:

> Thus, the *corpus delicti* requires, not only the burning of a building, but also that the burning was caused by a criminal agency. Or, as the rule is sometimes expressed, the *corpus delicti* consists of two elements: (1) that

[4] Arthur F. Curtis, *A Treatise on the Law of Arson* (Buffalo, N.Y.: Dennis & Co., Inc., 1963), p. 8. (Used with permission of Dennis & Co., Inc.)

the building in question burned; and (2) that it burned as the result of the willful and criminal act of some person.[5]

It is necessary to examine these two elements, one at a time. The first is "that the building in question burned." Just what does this mean? What constitutes a burning sufficient to prove arson? Several state supreme court decisions supply our answer.

The Illinois Supreme Court said, in *People* v. *Oliff*, ". . . there must be a wasting of the fibers of the wood no matter how small in extent."[6]

In *State* v. *Nielson*,[7] the Utah Supreme Court stated, ". . . The law is well established that a charring of fibers of a part of a building is all that is required to constitute a burning sufficient to make the crime of arson. Any charring is sufficient." The question was raised of whether an acoustical tile ceiling was a part of the building. The court continued, "We cannot see reason to make a distinction between a wood panel ceiling and an acoustical tile ceiling. Each is an integral part of the building."

Curtis summarizes this point as follows:

A house is not burned within the meaning of an accusation of arson when it is merely scorched, or smoked, or discolored by heat, but little more is required. The offense may be committed if any part of the wood be charred, if the fiber of the wood is destroyed, or its identity is changed.[8]

The second element needed to constitute the *corpus delicti* of arson is "that it burned as the result of the willful and criminal act of some person." Two very important points must be understood in connection with this second element. Evidence in criminal cases is of two types, direct and circumstantial. Direct evidence is eyewitness evidence. Because of their very nature, arson and related crimes are usually committed in secret. Rarely can an eyewitness be found to testify to the striking of the match. In the case of direct evidence, the witness has seen all material parts of the crime committed and recognizes or can identify the perpetrator. Direct evidence also comes about by a judicial confession. The subject admits guilt in court, either by entering a guilty plea or upon the witness stand. All other evidence is called circumstantial.

Consequently, circumstantial evidence must be relied upon, in most cases, to obtain a conviction in the crime of arson. The necessity for circumstantial proof of arson is discussed in Curtis, as cited in *State* v. *Kitowski*.[9] Curtis points out:

---

[5]*Ibid*.

[6]361 Ill. 237, 197 N.E. 777 (1935) at 780.

[7]474 P. 2d 725 (1970).

[8]Curtis, *The Law of Arson*, pp. 101–102.

[9]44 Wis. 2d 259, at 262.

Arson is one of those crimes which are peculiarly of secret preparation and commission; and it is very seldom that the prosecution can furnish testimony of an eye witness who observed the setting of the fire. The very nature of the crime is such that it becomes necessary for the State, in many, if not most, cases to rely upon circumstantial evidence to establish the guilt of the accused.[10]

The second point concerning the second element is one widely accepted by the courts as a method of establishing the *corpus delicti* of the crime of arson. In *People* v. *Wolf*,[11] the Supreme Court of Illinois followed the Missouri case, *State* v. *Cox*.[12] The court stated, ". . . separate, non-related, simultaneously burning fires are *prima facie* evidence of an incendiary origin." In other words, *when the fire department encounters two or more separate, nonrelated, simultaneously burning fires in a structure, it is evidence of an incendiary origin*. In *People* v. *Wolf*, the Illinois Supreme Court found, "Evidence of these separate fires was held admitted in proof of *corpus delicti*. The rule as above stated in *Corpus Juris* is also supported in *State* v. *Cox*, 264 Mo. 408."

The investigator may be able to establish a *prima facie* case of an incendiary fire. But regardless of the quality and amount of evidence produced for this purpose, the work is not finished. The competent investigator will eliminate all possible nonincendiary causes. If there remains any possibility that an accidental or providential cause can be established, the case can be lost.

Legal Decisions

The second element of the *corpus delicti* of the crime of arson is "that it burned as the result of the willful and criminal act of some person."[13] As indicated, most of the time this element must be proven with circumstantial evidence. One of the principal ways of connecting some person to the fire is by motive. An excellent example of the uses of circumstantial evidence is found in the Wisconsin case, *State* v. *Kitowski*.[14] The Wisconsin Supreme Court indicated that circumstantial evidence may be and often is stronger and more convincing than direct evidence. In this particular case, the circumstantial evidence was that the defendant and the woman owner of the farmhouse that burned visited the home of friends. While there, they had a violent quarrel. This quarrel resulted in the defendant's threat to burn the woman's dwelling "to the ground right now," after which he was physically ejected from the home he was visiting. About fifteen minutes later, the fire started. The defendant was

---

[10]Curtis, *The Law of Arson*, pp. 518–520.

[11]334 Ill. 218, 165 N.E. 619 at 622.

[12]264 Mo. 408, 175 S.W. 50.

[13]Curtis, *The Law of Arson*, p. 8.

[14]44 Wis. 2d 259 (1969).

observed in the immediate vicinity of the farm at this time. The next morning, the owner accused him of the arson, and he stood mute. Evidence was presented in the trial court that completely eliminated any possibility that the fire resulted from natural causes.

As can be seen, the circumstantial evidence established the motive, placed the defendant in the vicinity of the fire, and showed that when he was accused of the crime, he made no effort to deny the accusation. The jury considered this evidence to be sufficient to prove the defendant's guilt beyond any reasonable doubt. The Wisconsin Supreme Court affirmed the decision.

Searching the Fire Scene

If the fire investigator is going to determine the cause of a fire, he or she must search the fire scene. Locating the point or area of origin is the first step in establishing the cause of a fire. The question of when and how the fire investigator may search the fire scene may produce a difficult situation. The investigator is in the position of having to examine the scene before any determination can be made as to whether or not a crime was committed. But without a search warrant, any search of the premises may be illegal. Rarely does an investigator known the cause of a fire or explosion prior to conducting the investigation.

In 1967, the U.S. Supreme Court handed down two decisions that had a profound effect on the conducting of fire-prevention inspections. *Camara* v. *Municipal Court of the City and County of San Francisco*[15] was the first to be decided: San Francisco housing inspectors had made three unsuccessful efforts to secure Camara's consent for a warrantless inspection of the ground-floor residence he leased in an apartment building. Camara was charged with violating the San Francisco housing code for refusing to allow this inspection, allegedly in violation of the building's occupancy permit. Camara, upon his arrest and conviction, challenged the constitutionality of the housing code.

The Supreme Court held that the Fourth Amendment bars prosecution of a person who has refused to permit a warrantless code-enforcement inspection of his personal residence. It further pointed out that the basic purpose of the Fourth Amendment's prohibition of unreasonable searches and seizures is to safeguard the privacy and security of individuals against arbitrary invasions by government officials. The Court emphasized that any warrantless search of private property without consent is unreasonable. On the basis of the nature of the search, the Court has distinguished between regulatory, administrative, and criminal searches. *Camara* has now become the controlling decision in the United States in the area of administrative inspections and searches. Administrative inspections are those associated with a regulatory system for the protection of public health, safety, and morals.

[15] 387 U.S. 523, 87 S.Ct. 1727 (1967).

In *Camara*, the Court held that administrative searches are within the scope of the Fourth Amendment and require authorization by a search warrant. However, in the case of administrative searches, the magistrate could issue such a warrant on less than probable cause. In this decision, the Court set up a new type of warrant system for health- and safety-code inspections. It established a new "probable-cause" basis for warrants in the code-enforcement area. Warrants would not be dependent upon the inspector's belief that a particular dwelling violates the code, but rather upon the reasonableness of the enforcement agency's appraisal of conditions in the area as a whole. Search warrants, which are required in nonemergency situations, should normally be sought only after entry is refused.

The companion case, *See* v. *City of Seattle*,[16] involved a commercial warehouse owner. See refused to submit to a fire inspection of his locked warehouse. The inspection was conducted as part of a routine, periodic citywide canvass to obtain compliance with Seattle's fire code. After See refused the inspector access, he was arrested and charged with violating the fire code. Upon conviction, he too appealed on constitutional grounds. The Supreme Court held that the businessman, like the occupant of a residence, has the right to go about his business free from unreasonable searches. Official entries upon his private commercial property are unreasonable. Warrants are necessary and tolerable limitations on the right to enter upon and inspect commercial premises. The Court applied the same standards for search warrants in the *See* case that it had established in *Camara*.

A New Jersey appellate court upheld a warrantless search during a fire investigation in *State* v. *Vader*,[17] saying:

> The basic purpose of the Fourth Amendment is the protection of an individual's privacy and the security of his home. Here, the premises had been rendered uninhabitable by a fire. All utilities had been disconnected. No one was occupying the house, the doors and windows of which were broken. The fire was of suspicious origin and had resulted in the death of a child. Under these circumstances, the prompt, on-the-scene investigation of the fire by the authorities did not infringe on the defendant's right of privacy or the security of his home and was not a Fourth Amendment search requiring a search warrant.[18]

An appellate court in Oregon held that a warrantless entry would not invade a constitutionally protected interest in privacy if the owner or occupant of the burned premises had abandoned the property.[19]

[16] 387 U.S. 541, 87 S.Ct. 1737 (1967).
[17] 114 N.J. Super 260, 276 A. 2d 151 (1971).
[18] 276 A. 2d at 152.
[19] *State* v. *Felger*, 10 Or. App. 39, 526 P. 2d 611, 615 (1974).

Recently, the Michigan Supreme Court, in *People* v. *Tyler and Tompkins*,[20] held that warrantless searches are unconstitutional. The court ruled that, except under certain specific conditions, an administrative search warrant is required by fire authorities in the investigation of the cause and origin of a fire.

In this case, the fire broke out before midnight. Before the fire department left the scene, the fire chief discovered and seized two plastic containers, one of which was partially filled with a flammable liquid. The fire chief and a detective from the police department conferred before the firefighters left. The detective tried to take pictures of the interior of the building, but he was apparently hindered by darkness, steam, and smoke. By 4:00 A.M., the fire was extinguished and the fire department left. The premises were left unattended until 8:00 A.M., when the fire chief and an assistant chief returned briefly to survey the interior of the building. Then, for a second time, the building was left unattended. Sometime between 9:00 and 9:30 A.M., the officials again returned to the fire scene. At this time they discovered on the carpet of one room a thin linear burn that circled the room, went through a door, and continued down a stairway to an exit. They removed pieces of carpeting and wood containing the burn patterns—evidence that, at the trial, was admitted over objections.

It should be noted that in this case, the building was left unattended after the fire was extinguished. Important evidence was recovered by the investigators, but only upon their return nearly five hours later. The Michigan Supreme Court applied the standards established in *Camara*. It pointed out:

> The United States Supreme Court, in distinguishing between regulatory, administrative and criminal investigative searches, has provided a framework for achieving a workable balance between the need for investigation of the causes of fires and protection of the individual's right of privacy.
>
> If there are exigent circumstances, or the evidence is in plain view, no warrant is required. Nor is a warrant required for a prophylactic regulatory inspection of public places.
>
> If there has been a fire, the blaze extinguished and the firefighters have left the premises, a warrant is required to re-enter and search the premises, unless there is consent or the premises have been abandoned.[21]

The court indicated that, if arson was suspected, evidence seized under the following six conditions should be admissible:

1. Any evidence that is obtained while firefighters are on the premises putting out a fire is admissible under the "plain-view" doctrine.
2. As long as an emergency exists—which includes overhaul operations and an examination to prevent rekindles and abate hazards, such as leaking

[20] 399 Mich. 564, 250 N.W. 2d 467 (1977).
[21] 250 N.W. 2d at 467, 477.

gas lines and energized electrical lines—no warrant is required, and any evidence discovered should be admissible.

3. In certain situations where evidence may be destroyed or lost if immediate seizure does not occur, evidence thus obtained is admissible.

4. If consent of the owner or person in control of the premises is obtained, then evidence seized is admissible.

5. If the premises are so completely destroyed that there are no longer recognizable objects of personal property, the provisions of the Fourth Amendment do not apply concerning searches.

6. If the owner or occupant abandons the property, a search warrant is not necessary.

The court concluded:

Although post-fire searches made solely for the administrative purpose of determining the cause and source of the fire may properly occur under reasonable guidelines and with Camara's reduced standard of probable cause, where the investigation turns to the collection of criminal evidence, constitutional requirements cannot be subordinated to a sweeping grant of statutory authority.[22]

The U.S. Supreme Court affirmed the decision of the Michigan Supreme Court with modifications. The Court held:

A burning building clearly presents an exigency of sufficient proportions to render a warrantless entry "reasonable." Indeed, it would defy reason to suppose that firemen must secure a warrant or consent before entering a burning structure to put out the blaze. And once in a building for this purpose, firefighters may seize evidence of arson that is in plain view. *Coolidge* v. *New Hampshire*, 403 U.S. 443, 465-466. Thus, the Fourth and Fourteenth Amendments were not violated by the entry of the firemen to extinguish the fire at Tyler's Auction, nor by Chief See's removal of the two plastic containers of flammable liquid found on the floor of one of the showrooms.

Although the Michigan Supreme Court appears to have accepted this principle, its opinion may be read as holding that the exigency justifying a warrantless entry to fight a fire ends, and the need to get a warrant begins, with the dousing of the last flame. 399 Mich., at 579, 250 N.W. 2d, at 475. We think this view of the firefighting function is unrealistically narrow, however. Fire officials are charged not only with extinguishing fires, but with finding their causes. Prompt determination of the fire's origin may be necessary to prevent its recurrence, as through the detection of continuing dangers such as faulty wiring or a defective furnace. Immediate investiga-

---

[22] 250 N.W. 2d at 475.

tion may also be necessary to preserve evidence from intentional or accidental destruction. And, of course, the sooner the officials complete their duties, the less will be their subsequent interference with the privacy and the recovery efforts of the victims. For these reasons, officials need no warrant to remain in a building for a reasonable time to investigate the cause of a blaze after it has been extinguished. And if the warrantless entry to put out the fire and determine its cause is constitutional, the warrantless seizure of evidence while inspecting the premises for these purposes also is constitutional.[23]

Tyler and Tompkins argued that the Michigan Supreme Court was correct in holding that the departure by the fire officials from Tyler's Auction at 4:00 A.M. ended any authority they might have to conduct a warrantless search. They further argued that even if these fire officials had been entitled to remain in the building without a warrant to investigate the cause of the fire, their departure and reentry four hours later that morning required a warrant. The Michigan Supreme Court agreed with them; the U.S. Supreme Court did not. It modified the decision of the Michigan Supreme Court, holding:

> On the facts of this case, we do not believe that a warrant was necessary for the early morning re-entries on January 22. As the fire was being extinguished, Chief See and his assistants began their investigation, but visibility was severely hindered by darkness, steam, and smoke. Thus they departed at 4 A.M. and returned shortly after daylight to continue their investigation. Little purpose would have been served by their remaining in the building, except to remove any doubt about the legality of the warrantless search and seizure later that same morning. Under these circumstances, we find that the morning entries were no more than an actual continuation of the first, and the lack of a warrant thus did not invalidate the resulting seizure of evidence.
>
> The entries occurring after January 22, however, were clearly detached from the initial exigency and warrantless entry. Since all these searches were conducted without valid warrants and without consent, they were invalid under the Fourth and Fourteenth Amendments, and any evidence obtained as a result of these entries must, therefore, be excluded at the respondents' retrial. . . .
>
> In summation, we hold that an entry to fight a fire requires no warrant, and that once in the building, officials may remain there for a reasonable time to investigate the cause of the blaze. Thereafter, additional entries to investigate the cause of the fire must be made pursuant to the warrant procedures governing administrative searches. See *Camara, supra*, at 534– 539, *See* v. *City of Seattle, supra*, at 544–545; *Marshall* v. *Barlow's, Inc.*, *ante*, at 320–21. Evidence of arson discovered in the course of such investigations is admissible at trial, but if the investigating officials find probable

[23] 436 U.S. 499 at 509–510, 56 L. Ed. 2d at 498–499 (1978).

cause to believe that arson has occurred and require further access to gather evidence for a possible prosecution, they may obtain a warrant only upon a traditional showing of probable cause applicable to searches for evidence of crime. *United States* v. *Ventresca*, 380 U.S. 102.[24]

The key phrase in this decision is "reasonable time to investigate." Its interpretation will undoubtedly vary from state to state and court to court. However, in his decision, Mr. Justice Stewart did attempt to clarify the phrase in a footnote. He stated:

> The circumstances of particular fires and the role of firemen and investigating officials will vary widely. A fire in a single-family dwelling that clearly is extinguished at some identifiable time presents fewer complexities than those likely to attend a fire that spreads through a large apartment complex or that engulfs numerous buildings. In the latter situations, it may be necessary for officials—pursuing their duty both to extinguish the fire and to ascertain its origin—to remain on the scene for an extended period of time, repeatedly entering or re-entering the building or buildings, or portions thereof. In determining what constitutes a "reasonable time to investigate," appropriate recognition must be given to the exigencies that confront officials serving under these conditions, as well as the individuals' reasonable expectations of privacy.[25]

QUESTIONS

1. What is meant by the term *stare decisis*?
2. Describe the two types of precedents discussed in this chapter.
3. Discuss why a precedent is important in our system of justice.
4. List the three general causes of fires from the legal standpoint.
5. Discuss the two elements necessary for the *corpus delicti* of the crime of arson.
6. Discuss the significance of the *Camara* and *See* cases in connection with fire investigations.
7. What constitutes a burning sufficient to prove arson?
8. What is the significance of two or more separate, nonrelated, simultaneously burning fires in a structure?
9. What are the three kinds of searches recognized by the U.S. Supreme Court in the development of probable cause and warrant requirements?

[24]436 U.S. 499 at 511–12, 56 L. Ed. 2d at 499–500 (1978).
[25]436 U.S. 499 at 510, 56 L. Ed. 2d at 499 (1978).

Firefighters' Role
in Fire
Investigation

In most fire investigations, the investigator does not respond to the fire scene until summoned. He is unfamiliar with the conditions and circumstances under which the fire occurred. In addition, the persons and property involved are unknown. The investigator is only as good as his or her source of information. In this case, the members of the fire department on the scene must be this source.

What information should be available from the firefighters? The data required fall into three categories: (1) that type of information attainable or developed prior to the arrival on the fire scene, (2) the information available to the firefighters at the fire scene, and (3) the type of information that becomes available during overhaul and thereafter.

Before Arrival

The time of the receipt of the alarm is a most important factor. The time, day of the week, and data can all be significant in a fire investigation. If a business concern is involved, its personnel may assist in establishing how long the place was closed before the fire was discovered. Who was the last person to leave the premises? Did the fire occur during or after school hours? Was it during the daylight hours or at night? Were there likely to have been witnesses in the vicinity? What was the normal type of traffic in the area at that time?

Weather conditions can be another important factor. Rain, electrical storms, snow, sleet, ice, wind, humidity, and atmospheric inversion—all can have an effect upon the fire. The presence or absence of an electrical storm will determine whether lightning can be eliminated as the cause of the fire.[1]

Upon approaching the fire scene, the firefighter should observe any person

---

[1] *State* v. *Kitowski*, 44 Wis. 2d 259, 264 (1969).

or automobile hurriedly leaving the area. If the fire is blazing at all, it is most natural for people to be attracted by it. Careful mental notes about them or identifying characteristics of persons or vehicles leaving the scene may be of great value to the investigator. Some other conditions that should be noted are unusual road or street conditions, such as barricades slowing the progress of the response, or vehicles parked in such a manner as to create obstructions to the fire scene, hydrants, sprinkler connections, streets, or driveways.

The color of smoke and of flames can be a rather reliable indicator of the type of fire. Some of the more common of these indicators are:

1. Black smoke with deep red flames—petroleum products, tar, rubber, many plastics, etc.
2. Heavy brown smoke with deep red flames—nitrogen products
3. White smoke with bright white flames, reactive to water—magnesium
4. Black smoke with red and blue-green flames—asphalt shingles
5. Purple, violet, or lavender flames—potassium
6. Greenish-yellow flames—chlorine or manganese
7. Bright reddish-yellow flames, similar to fusee or railroad flares—calcium
8. Smoke of the usual color found in most fires that changes to yellow or grayish-yellow—usually indicative of a "backdraft" condition

The smoke and flames observed just prior to the arrival at the fire scene may be of great importance.

Upon Arrival

Upon arrival at the fire scene, several different conditions should be watched for. The existence of separate, nonrelated, simultaneously burning fires could be of great significance, as could the extent and area of the fire. Was forcible entry necessary to gain admission to the premises? Was entrance through unlocked doors? Was there evidence that the doors had been forced before the arrival of the fire department? Firefighters should look for marks on doors and windows that might indicate a forced entry, and notice whether tools used for this purpose have been discarded by the burglar. In one case, the fire department finally gained entrance to the office in a lumberyard fire. The firefighters found the safe wide open and its contents lying all over the floor. Since the fire was spreading rapidly, the firemen scooped up as many of the papers as possible, threw them back into the safe, and closed the safe. The building was destroyed, but the papers in the safe were salvaged. The burglarized safe was the first indication of an incendiary fire.

Account ledgers and other business records laid out on the desks in the office may be another indication of incendiarism. Destruction of business records may well be a motive for arson, particularly if the crime takes place just prior to an audit of the books. Embezzlement or a serious inventory short-age might be detected in the audit.

Had the fire spread, or was it spreading, with unusual speed? Was the fire more intense than one usually encountered in a similar occupancy? Did the fire behave naturally? Was there any unusual odor at the fire scene? Some arsonists have been known to use perfume, ammonia, or similar strong odorants to hide the smell of an accelerant. Were there any obstacles or obstructions to delay entry by fire-suppression forces?

The windows of the building should also be observed. Shades have been drawn and paper placed over windows to keep anyone from looking into the building. This gives the fire an opportunity to build up prior to discovery.

Upon entering the building, any unusual arrangement of the contents should be observed. There may be evidence of lack of stock or a substitution of stock. Sometimes new, expensive stock is removed and secondhand, old stock substituted for it. After a bad fire, there might not be enough left to tell the difference.

If the building has automatic sprinklers installed, other conditions must be verified and observed. Upon arrival, was the sprinkler system functioning prop-erly? Was it controlling the fire? If not, why not? Were the post-indicator valves open or closed? Were the O.S.&Y. valves open? Was the sprinkler alarm or water gong operating? Was the fire department Siamese connection operational? Any condition other than normal must be reported to the fire investigator. In addition, if the reason for the abnormality is known, it should be reported.

As fire-suppression forces proceed through the building, all fire doors should be noted. Were they open or closed? If open, were they blocked open or dis-abled? Was there any evidence that fire extinguishers or standpipe hose had been used to fight the fire? How did the fire behave when an attempt was made to extinguish it? Did it spread more rapidly, resist extinguishment, or "bite back"? Did it spread over a larger area when hit with fire-department hose streams? Did explosions result from attempts to extinguish the fire?

If the fire occurred at night, how were the occupants dressed? Did they have their best clothes on? Did they seem to be acting normally for the situation? Were they calm? excited? hysterical? If the fire did not seem to bother them, some explanation might be in order. Details such as this may be of great value later.

Who turned in the alarm? Was the person still at the scene upon arrival of the fire department? Is this person the same one who discovered the fire? If not, who told the person turning in the alarm to call the fire department? Was the owner or occupant at the fire scene? How did the owner or occupant learn about the fire, if he or she was not present at the time of the fire? Who notified the owner or occupant? When fires occur at night or after working hours, an explanation of why the owners or occupants are or are not present could be important.

Firesetters of various types frequently remain near the fire scene and await the arrival of the fire department. Sometimes, they even assist the firefighters in the extinguishment of the fire. The firefighter should be aware of familiar faces seen at several fires. It is also wise to note anyone who seems unduly excited or who is taking an unusual or abnormal interest in the fire. Conversations or remarks overheard can be important. All this information should be passed on to the investigator as soon as possible. (A detailed discussion of the various types of firesetters will be found in Chapter 10.)

The fire-pump operator must usually remain in the vicinity of the pumper and therefore has little freedom of movement. People watching the fire may wish to talk with or be near such a person. Conversations between spectators can be very revealing. Many pump operators have been able to give the investigator leads as a result of these observations.

During and After Overhaul

During overhaul, firefighters should refrain from cleaning out or disarranging the premises any more than is absolutely necessary, particularly in the area of the origin of the fire. The area of origin is usually the most heavily damaged by the fire, probably the most burned-out section of the building. Little additional damage is likely to result from not thoroughly cleaning up this area.

It is natural for the firefighters to clean up damaged debris during overhaul. It is usually thrown into one of two areas. The first is out the window. Most insurance policies provide for debris removal from the premises. However, traditionally the fire service has considered that debris removal was good public relations. True, the homeowner appreciates it. But his gratitude is nothing compared to that of the arsonist. He knows that the public-relations-conscious firefighter has buried the evidence of his crime under a pile of burned rubbish out in the yard.

The second area used for the disposal of badly burned or water-damaged debris is the heavily burned area of the origin of the fire. This is the area that the fire investigator must examine most carefully. The presence of this additional discarded and damaged debris creates a twofold problem: It will require that all the material be moved again; and it will create uncertainty as to exactly what was present at the time of the fire and what was discarded during overhaul.

The firefighter must be careful to check whether there has been any tampering with the utilities. Also, are all pets accounted for? Sometimes pets are removed prior to the fire. Have family heirlooms been removed from the premises? family pictures, wedding pictures, and wedding albums? The better clothing belonging to the family, particularly to the children, may have been removed.

Many things that may be observed by the firefighter are of interest to the investigator—multiple fires, odors, undue wood charring, uneven burning, inoperative sprinklers, tracks, footprints, evidence of a burglary, and so on. In addition, the firefighter may be the first to recognize trailers set between the multiple fires. These include string or cord that has been soaked in oil or wax; streamers of paper, toilet paper, or newspapers; excelsior; and other combustible and flammable materials. The use of trailers, chemicals, flammable liquids, and other types of material used to start fires will be discussed in Chapter 6.

If the firefighter discovers some objects that might be considered physical evidence, what is the procedure he should follow? First, if at all possible it should be left where it is if there is no danger that the object will be damaged or destroyed. A superior officer should be notified and the object pointed out. All should agree as to what it is and exactly where it was found. Its removal will be handled by the people conducting the investigation.

The firefighter, who may be on the scene long before the investigator arrives, must train to become the eyes of the investigator. Observation of unusual and unnatural conditions at the fire scene are the key to fire investigations.

QUESTIONS

1. Why is the color of smoke and flames important to the fire investigator?
2. What are some of the observations that should be made by the firefighter en route to the fire?
3. What are some of the observations that should be made by the firefighter upon arrival at the fire scene?
4. Why should firefighters refrain from cleaning out the premises any more than is absolutely necessary?
5. What should the firefighter look for during overhaul operations?

Searching the Fire Scene

The primary responsibility of every fire investigator is to determine the true cause of the fire. To this end, the fire scene must be carefully searched. The investigator's knowledge and understanding of fire behavior and the chemistry of fire are the very basis of the search. A good investigator will be prepared in two ways before conducting the fire-scene search: the first is in having the proper tools; the second, of equal importance, is in attitude. It is extremely important to keep an open mind when making an investigation. This means having no preconceived ideas as to the cause or origin of the fire. Objectivity is a cardinal characteristic of the good investigator.

Upon arrival at the scene, the investigator may find many "experts" with all kinds of opinions as to cause and origin. But this information must be accepted, however difficult it may be, with objectivity. Most firefighters, fire officers, and others at the scene have seen, heard, or experienced something that may be of value to the investigator. The ability to sift and analyze each bit of information will determine how objective the investigation becomes.

Some of the tools the investigator should have at the scene of the fire are:

1. Flashlight or lantern
2. Notebook, pen, pencils
3. Sketchpad or pad of graph paper for sketching the scene
4. Straight-edge ruler
5. 6-foot steel tape
6. 50- or 100-foot steel tape
7. Tape recorder
8. Photographic equipment
9. Device for measuring depth of char

A detailed listing of other equipment to be included in the investigator's kit will be found in Chapter 9.

The time of arrival at the fire scene should be recorded. As the investigator proceeds with the investigation, all observations are recorded. Many use both the notebook and tape recorder for this purpose.

Steps in Searching the Fire Scene

When making an investigation of a structural fire, many investigators will first make a cursory inspection of the structure from the outside. This outside survey will disclose any external origin of the fire. An external origin could be a communication of fire from another structure or an outside fire. Whenever possible, sketches should be made and photographs taken. These will establish and verify the exact conditions found at the scene.

The importance of reading and interpreting burn patterns cannot be overemphasized. As the investigator makes the external inspection of the fire building, these patterns must be carefully noted. If the structure is still fairly intact, a burn pattern coming out of certain windows and not others can be important. The interior areas in the vicinity of the windows with the external burn patterns are saying, "Look at me." If most or all of the roof or upper floors are intact, this will further aid the investigator in the search for the fire origin.

Other types of burn patterns of particular importance to the investigator may also be seen from the outside of the structure. For instance, the patterns in mobile-home fires should be observed carefully. Many mobile homes have exteriors of aluminum, which melts at approximately 1,200°F. The heat at the floor level in most structure fires will not normally exceed 500°–600°F. Any holes burned through the aluminum siding of a mobile home near the floor level should be carefully examined. Such an exterior hot spot may indicate where an interior hot spot is located, since the latter is usually found at a point directly below the external hot spot. The origin of the fire may be in the immediate vicinity.

In order to determine a point of origin, the investigator must first locate the area or room of origin. (This may have already been done by the fire department.) As a fire burns, it consumes combustible materials. Normally, the area where a fire originates will be exposed to the heat and flames for the longest period of time. Therefore, charring will be the heaviest and the destruction of the combustibles the greatest in this location. Thus, the origin is usually found in that part of the structure where the most severe damage has occurred. This may be a single room, a portion of a room, or an entire section of the structure.

To locate the place of origin, check the ceiling area first. Look for heavy damage to the upper parts of the room. Material destroyed by fire and severely damaged ceiling or roof rafters can be important. From these upper areas, trace the flow of heat downward. Follow the deeper charring from the ceiling down the wall to the lower areas of the room. That portion of the room where the destruction is most severe may indicate an area of origin.

BURN PATTERNS AND CHAR DEPTHS

As indicated in Chapter 2, the heat from a fire has a tendency to rise. As it rises, pyrolysis and the charring of combustible materials occur. As the pyrolytic process continues, the resulting char is deeper where the decomposition has been taking place the longest. This results in the formation of the inverted cone pattern, illustrated in Figure 2-4. The bottom of this cone pattern is the area of most interest to the investigator. Here is where the search for the point of origin begins.

Signs that assist in determining the low point of burning are charring at the bottom edge of shelving, bookcases, and tables. In addition, check for charring along molding, mopboards, and baseboards, under window sills, and on the bottom side of any other furniture with a flat surface exposed. The charred underside of any of these items will indicate fire from below. Always compare the depth of charring on the underside of a piece of furniture with the depth of char on its upper surfaces. Figure 5-1 illustrates this point. The char pattern

FIGURE 5-1. Char pattern on side of restaurant counter, indicating that fire burned upward to counter top.

found on the side of a restaurant counter was greater than that on the top of the counter. The fire department listed the origin of the fire as being a coffee maker on top of the counter, but the burn pattern indicated a lower point of origin. The laboratory examination of the coffee maker disclosed it to be in operable condition and functioning properly. The origin of the fire was established as in the hallway, beyond the counter.

Another indicator of the source of heat within a room or area can be a partially melted glass electric light bulb. As the heat develops in an area, it also heats the light bulb. The heat distorts the bulb in the direction of the heat. This is just another of the many small signs the knowledgeable fire investigator will observe.

Fires originating in upholstered furniture can create very definite patterns. If the chair or sofa is close to a wall, an outline of the piece of furniture may be etched in the wall after the fire. The same is true of the ceiling directly above the piece or pieces of furniture. Heavy charring is likely to be visible.

Some of the questions the investigator must consider are these: What is the condition of the chair or sofa? Is it badly charred on both back and front? Is it charred only on the back or front? Is the wall directly behind it heavily charred? Is there charring on the wall above the top of the piece of furniture? Has the ceiling heavy heat damage directly above, or is the damage uniform over the entire ceiling area?

Heavy damage directly behind the furniture and a uniform damage to the ceiling will usually indicate a slow buildup of heat within the furniture from a smoldering-type fire. This is typical of a cigarette fire. Heavy heat and fire damage on the wall above the top of the sofa or chair, together with heavy damage to the ceiling directly above, will indicate a fast-burning fire in the furniture. The damage to the furniture would not be deep-seated, as in the case of the smoldering fire, but would be found mainly on the outside of the cushions. Figure 5-2 illustrates these points.

Cigarettes burn readily, but the heat is dissipated rapidly unless the cigarette is well insulated. A cigarette lying on a counter top or wooden shelf will scorch the surface, but rarely more. Cigarettes on top of newsprint will scorch the paper. Various tests have indicated that a cigarette will burn, on the surface of the tobacco, at approximately 550°F when no draft is present (that is, when no one is "dragging" on it). Cigarettes are so constructed that they will burn entirely to ash while lying in an ash tray or on some surface. Cigars, by their construction, will tend to go out under the same conditions.

The shape of a cigarette will cause it to roll and bury itself in upholstered furniture. Here, it is insulated and the heat builds up. Heat sufficient to cause flaming combustion can develop under these conditions. The period of time necessary to evolve flaming combustion of upholstered furniture from a cigarette is approximately $1\frac{1}{2}$–$2\frac{1}{2}$ hours.

In contrast to fires from other sources in a room, which burn upholstered

FIGURE 5-2. Fire pattern on wall above and behind position of burned sofa. Note outline of sofa on both walls. (Racine Fire Department photo, by Firefighter Mike Cooper.)

furniture from the outside surface inward, the fire developing from a cigarette usually burns outward from inside the cushion. The heat buildup is slow, but a piece of furniture burned by a cigarette fire is often completely destroyed. The floor and floor covering (carpeting or the like) will be damaged only in the immediate area of the piece of furniture. The cushion springs of a chair or sofa, or innersprings in the case of a bed, are likely to be annealed by the slow-burning cigarette type of fire. This annealing causes the springs to flatten out, giving the appearance of a metal doughnut.

The U.S. Forest Products Laboratory has reported that most woods have a uniform rate of char development of about 1.54 inches per hour. This would produce about one inch of char in 40 to 45 minutes, when exposed to an average flaming temperature of 1,400°–1,600°F.[1] (Keep in mind that in a room with the fire load found in the dwelling type of occupancy, the flames rolling across the ceiling after a flashover will create a ceiling temperature in the neighborhood of

---

[1] However, the Forest Products Laboratory indicates that the initial charring rate is about 1/30 inch per minute during the first five minutes, and then is reduced to 1/50 inch per minute after 30 minutes. E. L. Schaffer, *Review of Information Related to the Charring Rate of Wood*, U.S. Forest Service Research Note, FPL-0145 (Madison, Wis.: U.S. Forest Products Laboratory, 1966), pp. 14–15.

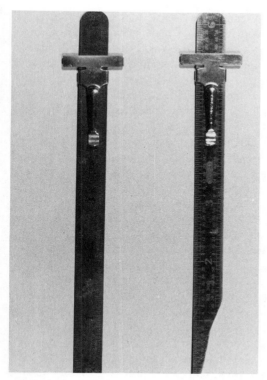

FIGURE 5-3. Six-inch flexible steel rule, showing
modification for use in measuring depth of charring.

1,400°–1,600°F.) The length of time that various wood surfaces have been
exposed to flaming combustion can be determined fairly accurately by the depth
of char. The investigator should have available some tool or instrument to
determine that depth. Some investigators use a tire-tread gauge; others modify
a six-inch flexible steel rule, as shown in Figure 5-3.

The combustion mechanism of wood is usually initiated by destructive distillation of the wood, or pyrolysis. This produces both combustible and incom-
bustible volatiles, in addition to the nonvolatile charcoal. The actual combustion
of the wood becomes the combustion of the products of this initial distillation.
Surface ignition and the charring of the surface of the wood occur simultaneously. As a result, ignition temperature and charring temperature are the same.[2]

Fire damage and charring must not be confused with heavy soot and other
products of combustion. The latter may be suspended in the air during the fire
and condensed into greasy film by the application of water spray from fire-
department hose streams.

[2]Schaffer, *Review of Information Related to the Charring Rate of Wood*, pp. 2–3.

As the fire and heat rise, they encounter obstructions such as the ceiling. This causes the horizontal movement referred to previously. The various products of combustion, particularly soot and other less volatile types, will move in the same manner. The horizontal movement continues until an obstruction, such as a wall, is encountered. Then they will begin to bank down. As this occurs, a trail of soot and greasy film is deposited. The heat that is accompanying these products of combustion will, in turn, have an effect upon everything it encounters.

Another pattern developed at this time that may be of value to the investigator is the "heat line" or "heat level." This is the line located on the walls of a room that indicates the distance from the ceiling to which the intense heat has banked or dropped down. It is indicated by blistering of the wall surface or discoloration of the wall or paneling, furniture, appliances, and fixtures. In many cases, these actually melt or char. Plastic fixtures such as lamps, telephones, radios, clocks, and wall tile are easily affected by the heat.

As the heat banks down after hitting obstructions to its horizontal movement, other patterns develop. The burn pattern or heat pattern from this action is usually rounded in shape rather than the traditional V pattern, or may even be a widened V. This pattern, however, may be modified by the characteristics of the environment of the area—shelving, soffits, drop panels above doorways or arches, even heating or air-conditioning units and ducts. Figure 5-7 is an example of this type of variation in a burn pattern.

Furniture, bookcases, and similar objects can also modify the traditional V pattern. The pattern may be bent or twisted. If no furniture is present because it was removed during overhaul, replacing it may reconstruct the whole pattern. Figure 5-2 illustrates this condition. The outline of the sofa is clearly visible. The wood-veneered paneling was protected below the top of the sofa and destroyed above it.

If the fire, heat, and products of combustion encounter any unprotected vertical openings as they flow across the ceiling, they will immediately seek a higher level. Some of the vertical openings permitting this upward communication are floor openings for stairways, elevators, and dumbwaiter shafts, pipe shafts and chases, openings around chimneys and flues, laundry or rubbish chutes, and ducts for heating, ventilating, and air conditioning. Many of these same types of openings, if horizontal and unprotected, will also permit communication of heat and fire. The investigator must be thoroughly familiar with all possible natural paths of fire communication and the subsequent burn patterns.

LOCATING THE POINT OR AREA OF ORIGIN

When the low point of burning has been determined, examine it carefully. Check for any residues indicating an accelerant or metal, and any material that would indicate a device used to start the fire. Many times this operation will necessitate the investigator's actually sifting the debris in the area of the origin.

Care is essential so that no evidence or possible evidence is overlooked. All possible evidentiary material must be carefully collected and preserved for the laboratory and/or the court. Details of this procedure will be discussed in Chapter 9.

In examing the area of origin, the investigator must also recognize any residues or materials that could have dropped from above, creating the low burn pattern. Such things as drapes, curtains, and burning panels of acoustical tile and ceiling are examples.

Flammable liquids flow to the lowest level on the floor surface. When they are ignited, the heat will travel upward. This will develop a pattern of charred surfaces. Burn patterns from flammable liquids on flat surfaces form in an ink-like blob outline. Patterns will be similar on the floor surface and on carpeting. Figures 5-4 and 5-5 illustrate these two patterns. Often, carpeting that appears charred all over may be carefully brushed with a broom after it has dried out, to bring out the distinct pattern of any accelerant used. In Figure 5-5, the couch was moved and the pattern revealed.

If the floor is charred, clear and clean the charred area very carefully. Examine and save materials of value. Check the type, pattern, and depth of

FIGURE 5-4. Flammable-liquid pattern on wooden floor surface. Note scorching where accelerant splattered and deeper charring where it was poured and along edges of wood flooring.

FIGURE 5-5. Flammable-liquid pattern on carpeting beneath sofa. Note coaster that indicates original position of sofa and its relation to burn pattern.

char at the probable point of origin. Compare the depth of floor charring with that of the ceiling directly above. At this point, the configuration of the room's environment becomes important. If the inverted-cone pattern has been altered, modified, or in any way changed, why?

The construction of the room and placement of its contents create the configuration of the room's environment. It cannot be too strongly emphasized that built-in cabinets, cupboards, soffits, bookcases, shelving, and even the placement of furniture affect the pattern the fire will develop. Windows, doorways, fans, and air-conditioning and heat ducts may all create unusual patterns. The investigator must be able to explain any unusual burning patterns found.

Figures 5-6 and 5-7 illustrate two types of unusual patterns. Figure 5-6 is the pattern developed from fire burning out of cans of flammable liquids sitting on a workbench. Note the patterns on the cement blocks from the heat of the cans and the fire coming out of them. Both the owner and the manager of the store claimed there had been no flammable liquids on the bench.

Distinctive burn patterns develop when fire and heat impinge upon noncombustible building materials such as concrete, cement blocks, or other masonry materials. As the fire builds up, soot and other tar-like pyrolytic products adhere to these materials. In the vicinity of any intense burning, the excessive heat will burn off these dark products of combustion, leaving white or very light-colored burn patterns. In the case of cement blocks, the heat can cause sufficient deterioration so that the investigator can feel the chalk-like surface. In the fire illustrated in Figure 5-6, the cement stuck to the fingers like soft chalk. The greater the heat applied, the more distinctive the pattern and bleaching process.

Figure 5-7 shows the unusual pattern developed in a supermarket fire. The liquor bottles on the top shelf were melted in one area but not on both sides. The air-conditioning duct shown at the top of the photo explained the unusual pattern. The air-conditioning unit was operating during the fire.

CONSTRUCTION AS APPLIED TO FIRE SPREAD

A thorough knowledge of building construction is of utmost importance to the fire investigator. One type of construction that can cause problems with the communication of fire is balloon frame. In balloon-frame construction, studs are usually 24 or more feet in length. This permits a clear vertical opening between the studs from the basement to the upper floors and the attic. Modification of

FIGURE 5-6. Burn patterns of flammable liquids in containers on bench. Note heat patterns at side of cans and above. Concrete-block wall had chalk-like appearance and consistency after fire.

this type of construction by inserting fire stops is the exception rather than the rule.

The investigator must be aware that when a fire has gained entrance to the inner structure of a balloon-frame building, its spread is unlimited. It may communicate throughout the entire structure. Most of the older two-story frame houses found in this country are of balloon-frame construction.

Another very common type of building construction is the brick-ordinary wood joisted structure, found along "Main Street" in most cities and towns. It has usually been modified and remodeled many times. One of the principal characteristics of this "ordinary construction" is the number of concealed spaces for fire to develop and spread.

Many older brick-ordinary joisted two- and three-story buildings have wood lath and plaster partition walls of balloon construction. If fire destroys in some way the integrity of the wall and gets inside, it will communicate upward immediately. In addition, fire traveling within this type of construction can move horizontally, perhaps through the channels between the floor joists or across the attic. If any opening exists within the wall, such as duct or pipe openings, fire will easily communicate into another room or area. Within the balloon-type

FIGURE 5-7. Burn pattern as indicated by melting of bottles on top shelf, caused by air coming out of the air-conditioning-system duct at the top of the picture. Note that the bottles were not melted along the entire top shelf, but only under the ducts. Origin of the fire was to the left of the picture.

constructed walls, fire may also spread downward as pieces of burning materials drop within the wall. All this may give the appearance of separate, nonrelated fires, but before such an origin is established, all natural routes of fire spread must be eliminated.

Whenever the downward fire or burn pattern is equal to that of the upward burn pattern, an explanation is most important. Anyplace within a structure where a downward burn pattern is found, it should be carefully examined and its cause determined. It is always possible that some combustible material has been affixed to or set against the wall. However, normal downward communication of fire can occur between stud channels in balloon constructions, as indicated above.

In any room within a structure, the doorway is the area where the heaviest foot travel occurs. This results in the normal wearing down of the floor surface at this point. Any flammable liquids on a hard surface will tend to puddle in these depressions or worn spots. And this tendency to puddle can be found even in newly constructed buildings. Figure 5-8 shows the charring from such a puddle in the doorway leading into a hall.

FIGURE 5-8. Burn pattern on hardwood flooring from the puddling of flammable liquid.

Check the bottom edges of doors and note any severe charring there. Normally, the bottom edge of a door will be uncharred during a structural fire; damage to this portion of the door will indicate an unnatural condition. It may be necessary to remove the hingepins and remove the door to conduct this examination. Figures 5-9 and 5-10 illustrate the charring from puddled flammable liquids under a door.

Regardless of the amount of severe destruction to the upper levels, including the ceiling and roof, the charring of floors is of major concern. In the event a floor is burned through, it may indicate that the fire originated from below. When charring is severe and a hole is found in the floor, examine the hole carefully. Try to determine the direction of fire travel. Many times, this may be done by observing the slant or angle at the edge of the hole. Figure 5-11 shows the point of origin of a high school fire. The fire started in an electrical junction box against a floor joist in the basement. It communicated up, burning a hole in the floor of a classroom. This fire will be discussed in more detail in Chapter 6.

FIGURE 5-9. Door to sacristy of church.

FIGURE 5-10. View of burn pattern on bottom of door to sacristy. Over the years, priests stepping on the marble threshold had created indentations, in which accelerant collected. Note burn pattern, which corresponds to position of indentations.

When a floor is burned from the top down, the charred surface will extend farther out from around the hole in the floor. In the same manner, a fire originating below the floor surface will show greater charring on the bottom side of the flooring adjacent to the hole. Figures 5-12 and 5-13 further explain these types of char patterns.

Other types of holes in floors include those burned through with round, almost vertical sides. This type of burn pattern may indicate the presence of heavy oil drippings. Over a period of years, oil from machinery such as lathes, boring equipment, sewing machines, refrigerators, washers, and similar stationary equipment may collect on the floor in one spot and soak through the flooring. Fire will attack this oil-soaked surface, and the result is this type of burn pattern.

Insects, such as termites, can reduce the mass of material in wood. Fire may attack the wood thus infested and burn holes through faster than would normally be expected. This applies to floor joists and roof rafters as well as flooring. Dry rot is another way wood may become deteriorated.

FIGURE 5-11. Fire started from electrical arcing in junction box against floor joist is indicated by black burn pattern on limestone basement wall. Fire communicated through wooden floor to classroom above.

Heavy use of a particular area or spot for foot traffic will wear flooring. Besides doorways areas such as the floor immediately in front of the bottom step of a stairs, or at the top of the stairs, are examples. The marble threshold in the doorway of the church where the door in Figure 5-10 was removed illustrates this point. The priests had been entering the church from the rectory through this doorway for many years. Two indentations had been worn on the threshold. The flammable liquid used as an accelerant in this fire puddled in these indentations. In the case of a combustible floor, the amount of wood available for burning may be much less and burn through faster.

Loosely or tightly constructed floors will also determine the type of burning pattern. Flammable liquid will penetrate loosely constructed floors and may result in burn patterns on the underside, as seen in Figure 5-14. Here, a loose-fitting floor permitted the flammable liquid to penetrate between two pieces of plywood subflooring and flow along and down the floor joist. The fire followed the same route and charred the underside of the subflooring and the floor joists in the basement. The point of origin of this fire was on the first floor, but the lowest point of burning was in the basement. In this case, the fire communicated from the top surface of the floor downward, following the path of the flammable liquid. This condition will be found in old tongue-and-groove flooring that has become worn and loosened over the years.

Openings under eaves into attics were very common for many years, until

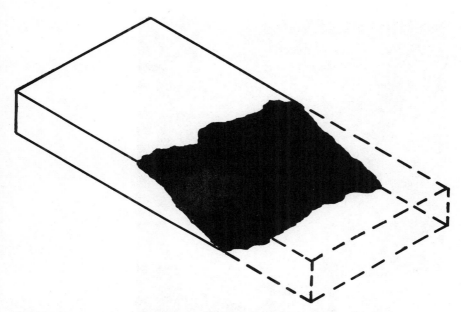

FIGURE 5-12. Pattern of hole in floor when fire burns from top down. Note greater area of charring on top of floor than on bottom.

the decrease in energy supplies necessitated better insulation. The wide-open type of attic could easily create air currents that developed unusual burn patterns in fires. The fire investigator may encounter this condition even today. In one case, the fire department pulled the second floor ceiling and dropped a pigeon nest onto the floor. This confirmed that the attic was subject to strong wind drafts, accounting for the resulting burn pattern.

FIGURE 5-13. Pattern of hole in floor when fire burns from below upward. Note greater area of charring on lower side of floor.

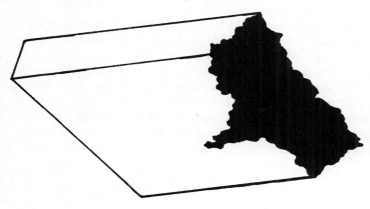

FIGURE 5-14. Burn pattern of accelerant that penetrated plywood floor at seams. Note how fire scorched floor joist as well as
bottom side of subflooring.

Searching Other Areas

The investigator must carefully check the windows and doors of the fire building. Was forcible entry necessary, or were the doors unlocked? Who made
forcible entry? Figure 5-15 shows a broken industrial-type window. Apparently,
the fire department forced the window and opened the latch to gain entry.
However, the investigator carefully checked the interior, especially the window
and window sill. Figure 5-16, was taken from the inside of the same window.
Note the smoke and soot stains on that portion of the window that extends
outside. This would indicate that the smoke and other products of combustion
flowed out of the window at this point. In addition, the investigator found
that the glass lying on the box on the left side of the window ledge came from
the broken window. This glass was unstained; the glass still in the window was
stained. An inventory disclosed that the fire was arson to cover the crime of
burglary.

Always check the garage and any detached sheds on the property. Sometimes a shed or detached garage is used to store furniture, clothing, and other valuables before or during the fire. Those personal items missing from the house may have been conveniently moved. If any evidence is discovered in a detached building, consideration should be given to the ramifications of the search procedures set forth in *Michigan* v. *Tyler*.

Many fire investigators are very apprehensive of any fire that originates in a clothing closet. Always check all closets. If the closet has been burned by the fire, look for common metal clothing hangers. These do not burn and should be found in the debris. The absence of hangers can indicate the absence of clothing, raising the question, Why was there no clothing in the clothes closet? Few natural or accidental fires occur in closets; most of them are incendiary or the result of children's playing with matches. Also check for clothing in dresser

FIGURE 5-15. Broken window in office of industrial plant. Note smoke on upper portion of window and none on lower.

drawers. The absence of clothing or of the clothing of a particular person must be explained.

Another item that should not be overlooked during the investigation is the food stock. Check the pantry and cabinets. How well stocked with food was the dwelling or apartment? Check the refrigerator. An empty refrigerator could indicate that the occupants did not anticipate occupying the premises after the fire. How about the good china or crystal? Is it still in the cabinet, or was the cabinet empty at the time of the fire?

Were there any pets in the household? Where were they during the fire? Are they missing? Other personal items that must be accounted for are wedding pictures, collections, the family Bible, and similar personal property. Few people deliberately destroy cherished and highly valued personal property. Some good examples are the hunter's guns, the fisherman's rods, reels, and tackle, the antique collector's prized items.

FIGURE 5-16. Same window, seen from inside building. Glass on box on left side of window sill was unstained, indicating it had been broken out of window prior to fire. Fire was arson to cover burglary.

Other Burn Patterns

Burn patterns from fires are a kind of sign language. The fire investigator must be able to recognize and interpret them correctly. The traditional V or inverted-cone pattern is probably the most common and easily recognized. Some of the V patterns are very sharp in shape; others spread out, making a wide V. The patterns vary as they move outside a room or structure through an opening. The width of the opening at its top can affect the pattern. Figure 5-17 illustrates the wide pattern from the doorway of a garage. The fire rose upward, broke the bedroom window, and communicated into the bedroom. The children sleeping in the bedroom died. Note the additional char pattern coming out of the window to the right. This type is frequently referred to as a "fan" pattern. It can occur as the result of the fire's moving out of the wider opening, or if the material burning at the point of origin is further away from a wall.

A fairly narrow V pattern can be seen in Figure 5-18. This incendiary fire was started on the back porch, ground floor. It spread around the corner of the house and moved upward along the clapboard siding.

Other patterns include modifications of the V or fan-shaped burn. Some, such as those that would be found in the channels between studs, rise straight

FIGURE 5-17. Fan or wide burn pattern rising out of garage. Fire communicated to bedroom through window.

FIGURE 5-18. Narrow V pattern resulting from fire spreading around side of building. Note pattern of flammable liquid applied to siding of upper porch.

up. Others are partial V or partial fan patterns. If there is no apparent reason for the type of pattern found, the investigator must determine what caused it or why it appears the way it does. If furniture has been moved, its replacement may assist in reconstructing the pattern or establishing the reason for it.

Burn patterns that rise in a normal manner and then begin to descend or move downward should be carefully examined. Patterns may rise at a sharp angle in a particular area and in this way indicate specific movement or direction of the fire. In one case, a closed door at the top of an open stairway was completely destroyed by the fire. However, a distinctive burn pattern was developed in the deeper charring at the bottom of the door jamb. It inclined upward at an angle of between 45° and 50°. The fire had built up from the first floor, lapping over the landing at the top of the stairs. The fire was therefore directed against the lower portion of the door. As the door was consumed, the lower part of the door jamb sustained the greatest amount of char. This enabled the investigator to trace the fire back to its origin.

Flammable liquids may be effectively applied even on the vertical surface of walls. They will create a recognizable burn pattern under these conditions. Figure 5-19 is an excellent example of severe charring throughout a stairwell.

FIGURE 5-19. Distinctive pattern from flammable liquid sprayed on wall and door. Wall was of hardwood wainscoting. All other wooden surfaces in stairwell were evenly burned.

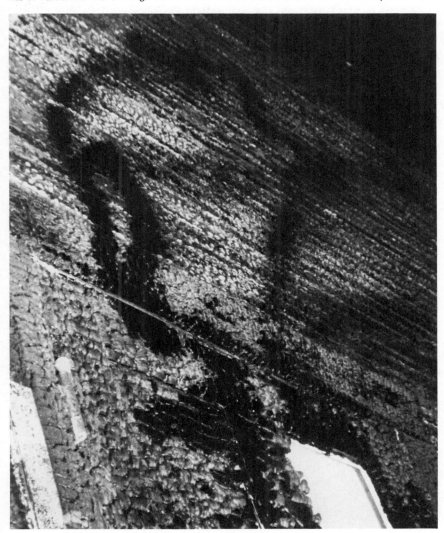

This stairwell was located between two storefronts. It extended from the sidewalk level to the second floor and was completely lined with hardwood wainscoting. The char pattern was extremely even throughout the entire stairwell, except for a very distinct pattern. This pattern was found on the top of the door that was at the bottom of the stairs and on the hardwood wainscot wall directly above the door. This burn pattern was well defined and a severe, but even, char about 3/16 inch deeper than the overall char pattern. This configuration was the result of an accelerant's being applied to the door and the wainscoting directly above. It is clearly visible in the photograph.

Note a similar pattern on the second-floor porch siding in Figure 5-18. This too was the result of the application of an accelerant.

The investigator surveying the fire scene must remember to observe the unusual, or the absence of the usual—something that is different from normal, something that is out of place or foreign to the particular occupancy. Any deviation from what would normally be expected must be noted. Many times, everything is right in front of the investigator; but he or she must be able to see it and to recognize it. A fire investigator, like a fire inspector, must be familiar with all types of construction and all types of occupancies. A fire investigation can challenge the investigator's entire lifetime experience and knowledge.

QUESTIONS

1. What is the primary responsibility of every fire investigator?
2. What is the meaning of the inverted-cone pattern?
3. Why is the char pattern usually deeper in the area of the origin of the fire?
4. What is the period of time usually necessary for flaming combustion to develop in upholstered furniture from a cigarette?
5. What is the significance of charring on the underside of a table, desk, or shelf?
6. What is meant by "configuration of the room's environment"?
7. Why is the low point of burning so important to the fire investigator?
8. What is the significance of a charred bottom edge of a door?
9. How may the direction of fire be determined from holes burned in the floor?

Fire Causes

After the point of origin has been established, the next step is to determine the true cause of the fire. Often, when the exact point of origin is established, the cause of the fire becomes obvious. This is particularly true when it is the only possible cause in the area of the origin.

As was indicated in Chapter 3, the courts recognize three basic causes of fire:

1. An act of God or providential origin
2. An accidental origin
3. An incendiary origin

When attempting to establish the cause of the fire, the investigator must keep in mind the three legs of the fire triangle—fuel, heat, and oxygen. A fire cause involves the two controllable conditions, the fuel supply and the heat source; the oxygen supply is always present.

Some types of fuel supplies available are:

Flammable and combustible liquids
Combustible solids and metals
Combustible gases
Combustible dusts

Heat sources include:

Open flames	Chemical reactions:
Hot surfaces	Polymerization
Electricity	Oxidizing agents
Mechanical sparks	Compression of gases

Lightning   Static electricity
Friction   Spontaneous heating

Natural Fire Causes (Acts of God)

Some of the providential causes of fire are lightning, spontaneous heating, and the rays of the sun.

LIGHTNING

Lightning strikes will result in fires only when sufficient heat is produced to ignite combustible materials. Fire will not occur when insufficient heat is produced; however, damage may still be extensive. There are two general types of lightning bolts, the "hot" bolt and the "cold" bolt. The "hot" bolt is one of longer duration and will ignite combustibles. The "cold" bolt, of shorter duration, has a tendency to shatter and splinter the building materials or even literally blow a structure apart.

Lightning, a form of static electricity, is an electrical current of great magnitude, producing tremendous amperage and voltage. Lightning strikes objects such as trees and structures because they contain materials that are better electrical conductors than the air. It enters buildings and structures by several means. It may enter by striking an object that is an electrical conductor and extends above or out of the building. Usually these objects are TV antenna, air-conditioning unit, CB antenna, or a part of the building, such as a cupola, tower, or chimney. The lightning may also strike the structure directly. It may hit power, telephone, and other transmission lines or poles, then follow the conductor into the building, or hit a tree or other tall object and leap over to the structure.

When lightning comes into a building by following the wires from a pole, the fuses in the fusebox may be blown. Figure 6-1 shows an electrical fuse subjected to the heavy current of a lightning bolt. The lightning set fire to the house and caused considerable damage.

One of the effects of a direct lightning hit on a structure can be a flashover between the electrical circuitry and the ground, which causes severe damage to any electrical equipment on the circuit. Lightning damage by a near miss can cause an induced line surge. This creates a greater problem, because the effect may be less obvious. When a surge is carried by a power line, it can affect everything connected to the line.

Always check the main switch and circuit breakers. If they were closed and carrying current at the time of the lightning strike, their contact points could be welded together by the tremendous heat generated. Seeing the melted copper and being unable to open the switches is evidence of this condition. Another indication of a line surge is damaged or burned insulation of a motor winding, usually found near or in the motor terminal box. An open breaker or switch can give a great deal of protection from a line surge. Remember that a motor that is operating can be damaged by a line surge from lightning. One that is not operating will probably sustain no damage. Therefore, if a motor was not running but has been damaged, the investigator should view with suspicion the story that lightning did the damage.

The investigator may be told that lightning hit a barn and that the barn exploded into flames. This is usually the result of the lightning's striking the barn in the hay-mow area. When it hits the steel rail used for moving hay within the mow, the rail becomes white-hot. The force of the lightning strike shakes the barn sufficiently to place the combustible dusts in suspension. The combination of the hot steel and combustible dust results in a flash fire or dust explosion.

SPONTANEOUS HEATING AND IGNITION

Spontaneous heating is the result of the slow oxidation of a combustible material. If this slow oxidation, with its heat production (exothermic decomposition), continues long enough and the heat is confined, the ignition temperature of the material will be reached. At this point, spontaneous ignition can take place. This has been defined as the burning of a combustible material without an external source of ignition.

Most types of spontaneous heating or combustion encountered by the fire investigator will fall into two general categories. The first comprises those easily oxidized combustible materials that produce sufficient exothermic decomposition at ordinary temperatures to reach their ignition points. These include such

FIGURE 6-1. Electrical fuse subjected to the heavy current of a lightning strike, a "hot" bolt that did considerable fire damage to the house.

substances as vegetable and animal oils, pyrophoric metals, charcoal, coal, and metallic sulphides. The second category is of those organic agricultural products subject to fermentation, such as hay and grains.

The fire investigator is going to repeatedly encounter suggestions that fires have started by spontaneous heating or combustion of "oily rags." Only animal and vegetable oils are subject to spontaneous heating. These are the oils known chemically as fatty acids. Oils from the methane series, usually derived from crude or mineral oils, are stable and not subject to spontaneous heating. Therefore, the story of oily rags causing a fire may be valid only if the oil involved is animal or vegetable. The fire investigator must use care not to be led into believing that any type of oily rags is subject to spontaneous heating.

In one case, the safety director of a large food-store chain encountered a spontaneous-heating incident in one of this stores. Four half-gallon containers of corn oil were broken in the service area. The employees used about 20 pounds of a kitty litter to absorb the oil, then placed the oil-soaked material in a cardboard box. The next morning, a strong, foul odor was detected coming from the box. It was discovered that the box was very hot. The mixture was removed from the store. The safety director (a fire-science student) began a series of experiments with vegetable oils and this particular kitty litter. He found that the combination could cause ignition with 24 hours. This store chain had suffered six large-loss fires that had destroyed as many stores. After the experimentation, the cause of three of these fires had been established.

The fire investigator in rural areas is likely to encounter hay-mow fires from spontaneous heating. Charcoal briquets that have been water soaked, because they are subject to spontaneous heating, can be the cause of fires in garages, where the wet charcoal is often stored.

RAYS OF THE SUN

Fires have also been caused by the rays of the sun, mostly as the result of the sun shining through older window panes containing "bubbles." More recently, there have been reports of the sun's rays coming through a window and hitting a concave mirror, concentrating the light on combustible materials and igniting them.

Accidental Fire Causes

CIGARETTES

One of the most common causes of fire, according to available statistics, is related to smoking. Cigarette fires were discussed in the preceding chapter, but several additional points should be brought out. Today, a great deal of upholstered furniture is being padded with soft urethane material. This type of plastic reacts to cigarettes differently from foam rubber and from cotton batting, previously in general use. Confined within the upholstery of cotton or rubber, the cigarette can cause ignition. However, a cigarette will not usually cause a fire in the urethane type of upholstery. The urethane tends to shrink away from the heat of the lighted cigarette, creating a hole through the urethane cushion or upholstery. However, many pieces of furniture still have cotton batting in the arms and back cushions. The investigator must be careful to determine the materials contained in the upholstery.

Urethane types of plastic, both soft and rigid, will burn if heated sufficiently, but the heat from the lighted end of a cigarette usually does not develop sufficient heat. When these plastics do burn, they do so rapidly, generating a great deal of heat, toxic vapors, smoke, and tar-like products. Figure 6-2 shows the destruction of a wood-lath and plaster wall behind an old sofa stuffed with cotton batting. A cigarette dropped in the sofa was believed to have caused the fire.

Many after-hours tavern and restaurant fires have been caused by the careless disposal of smoking materials. The fire in a combination pizza parlor and tavern originated on the second shelf of the cart shown in Figure 6-3. Soiled napkins had apparently been thrown on top of some ash trays containing "live"

FIGURE 6-2. The destruction of a wood-lath and plaster wall directly behind an old sofa that was stuffed with cotton batting and had burned very slowly.

FIGURE 6-3. An after-hours tavern fire originated on the second shelf of the cart seen here. Note the ashes of the burned linen napkins that had been on the second shelf. Under them were the remains of heavy ash trays and burned cigarettes.

cigarettes at closing time. The cart was positioned in a hall next to the kitchen doorway. The fire burned through the veneer paneling and throughout the hallway, consuming the veneer and communicating to the main area of the restaurant. Figure 6-4 is a photograph of the spot in the hallway where the serving cart had been left. Note the charring of the studs at a height even with the second shelf on the cart. In the background is the side of the counter, also veneered, where the fire spread and developed its pattern. This is the same counter seen in Figure 5-1.

ELECTRICAL FIRE CAUSES

The source of almost all electrical fires can be classified as originating from three conditions: (1) arcing, (2) sparking, and (3) overheating.

Arcing · If a short circuit or break in an electrical conductor occurs, the electric current tries to continue to flow in the open space created. As the current jumps across this open space, it creates an arc. The intensity of the arc depends upon the voltage and amount of current flowing.

Arcing on a small scale can occur every time an electric switch is opened. Its intensity, again, depends upon the amount of current flowing. If the atmosphere in which the switch is opened contains flammable vapors, flammable gases, or combustible dusts and the mixture is within the flammable range, an explosion or fire can occur when the switch is opened. An electric arc can create a temperature of as much as 7,000°F. The intense heat of arcing will cause a fusing or beading of the conductors.

The temperature necessary for fire to burn through a conductor is in the neighborhood of 2,000°F. Ettling of TFI Services reported on some recent experiments with the reaction of heat on electrical wires. These experiments were conducted using copper wire from 20 to 14 AWG (single conductor) and 20 and 18 AWG (stranded). He reported:

> A relatively cool fire in a building will not melt copper wires (1,083°C or 1,981°F). However, fires in general will cause a succession of changes in

FIGURE 6-4. Cart in Figure 6-3 had been positioned in hallway next to partition wall. Note excessive charring on studs behind wall paneling at the approximate height of second shelf of the cart. Fire communicated from this area up the side of the counter and over the counter along the wall at the rear of the tavern area.

copper. First, there will be oxidation and discoloration of the copper. The surface remains essentially smooth, and the longitudinal striations from the drawing process are still recognizable. As the melting point is reached, there is evolution of gases from the copper. This causes internal cavities and small blisters on the surface. As the surface starts to melt, there will be some distortion plus blistering, causing the longitudinal striations from the drawing process to be obliterated. As heating continues, there is free flow of melted copper along the surface of the wire. There is a tendency for the center of the wire to remain as an unmelted core. The flow of melted copper produces beads with thinner core areas between the beads. The core usually shows at the end of the wire as a point. Sometimes the ends are rounded with beads, generally having a rough surface. Sustained heating at just over the melting temperature sometimes gives smooth surfaces, because the gases are all boiled out and copper has a change to flow over the surface. More commonly, fire-melted conductors show rough surfaces. The interactions of oxidizing or reducing conditions or reheating after temporary cooling are not clear but evidently affect the appearance of the melted copper. If the fire is cooled at any stage in the preceding description, the copper will solidify and remain in that condition for examination after the fire.

When stranded conductors are fire-melted, copper normally will flow among the strands and fuse them into a solid wire with a badly distorted surface. This can vary, though, from slight incipient fusion to nearly complete fusion, short of dropping off. The two or three wires found in an NM cable will fuse together if they are in contact. The characteristic appearance of the surface will be essentially the same as for a single conductor, but two or three will be partially fused.

The effects of fire melting will be seen wherever the fire was hot enough. This might include several separate regions on a given set of wires. The hot spots will likely be between ceiling joists where the wires are fully exposed to full heat. Where wires go through framing members and especially through insulation, the melting temperature is normally not reached.[1]

Frequently, the wires are in a conduit, either thin-wall or pipe, or within a metal-clad cable (BX). Arcing within one of these conduits or metal-clad cables can burn a hole through the metal. When this happens, lips are usually formed around the hole on the opposite side from which the arcing occurred. Figure 6-5 shows such a hole burned through a BX cable from arcing on the inside. The lip formed on the edge of the hole can be seen in the photo. Figure 6-6 shows the same BX after being cut open and pulled apart. The bead on the one wire and indentation on the other conductor indicate the location of the short circuit within the BX.

[1] Bruce V. Ettling, "Electrical Wiring in Building Fires," *The Fire and Arson Investigator*, 29, No. 4 (April–June 1979), 10–19.

Figures 6-7 and 6-8 show the holes burned through a conduit. This was not a thin-walled conduit; it was one-inch galvanized-steel pipe with walls 3/16-inch thick. The lips formed are clearly visible. This electrical conduit provided the service for a junior high school, running through the basement of the senior high school. A series of shorts, with heavy arcing, occurred within a junction box affixed to the floor joist in the basement of the senior high school. The safety

FIGURE 6-5. Metal-clad cable (BX) with hole burned through from internal arcing. Note lip around hole in cable. Lips of this type are formed around hole on opposite side from where arcing occurred.

FIGURE 6-6. Same metal-clad cable cut in half to illustrate broken conductor and signs of arcing. Note bead formed on one end of broken conductor and indentation on the other end, at the point where the two came together, resulting in the arcing.

FIGURE 6-7. Holes with lips burned through a one-inch galvanized-steel pipe conduit. Note heavy lips around holes and where arcing burned a hole on the edge of the conduit inside the junction box.

FIGURE 6-8. Closeup of conduit and holes, showing lips formed on side opposite of arcing. Note the thickness of pipe the arcing burned through.

devices for electrical service function on the heat produced by the overload. In this case, the fuses were located approximately 150 feet from the junction box. The size of the wiring called for 70-ampere protection. The fuses were rated at 200 amperes. The arcing burned holes through both the conduit pipe and the junction box in several places. This, in turn, ignited the floor joists.

Figure 6-9 shows one of the connectors from the box and the beaded wires in it. Two of the connectors were recovered intact. Small pieces of the third, which had been literally blown apart by the arcing, were found in the debris. A combination of excessive fusing and the long distance from the point of shorting to the fuse block permitted this fire to occur.

Remember, it is the heat transmitted through the conductor to the safety device that causes it to function. It is heat that melts the fuse; it is heat that causes the circuit breaker to trip. If a fuse box or the circuit breakers are installed on an exterior wall, extreme temperatures may affect the speed at which the safety devices operate. In very cold weather, the action of the fuse or circuit breaker may be slowed down enough to permit a fire to start.

Always check the fuses and circuit breakers at the fire scene. Overloading or partial short circuits will melt the fuse band. A full short circuit will blacken and even distort the fuse face. Figure 6-10 shows a fuse with a face badly distorted and blackened by a full short circuit.

Always determine if the safety devices have been deliberately tampered with or bypassed. Most investigators are aware of the practice of placing a penny behind a fuse. Another method of bypassing a fuse is seen in Figure 6-11. A piece of the metal, forming the thread by which the fuse is screwed into the fuse block, is cut, then folded over onto the center pole. This will effectively render the fuse useless as a safety device. The potential fire condition created by this practice could be deliberate. Only an examination of the fuses in the fuse block will reveal this condition.

FIGURE 6-9. One of the three connectors with beaded wires still in it. Note heavy beading from heat and arcing.

FIGURE 6-10. Fuse with distorted face from a full
short circuit.

Short circuits occurring in or next to lamp sockets will usually be indicated
by a fusing of metal, blackening of the socket, or even splattering of small
globules of molten metal.

Another seemingly innocent electrical fitting that can cause a serious problem is the replacement of the male plug with the bayonet-type fitting for the wire.
These carry a UL label. The bayonets have very small, needle-type points that
penetrate the insulation of the wire to make contact in the conductor. It is
through these points that the electrical current must flow. Any movement of the

FIGURE 6-11. Fuse tampered with and modified to
bypass safety feature.

wire can create a poor connection, with the resultant heating or arcing. Overheating or a series of arcing sparks could easily ignite any nearby combustibles.

Sparks · Sparks most commonly occur as the result of electric (arc) welding operations or from the arcing of short circuits. The tremendous heat created during the electrical welding process creates a bond between the electrode and the material being bonded. This also creates many sparks and glowing globules of hot metal. If combustible materials are in the vicinity, the sparks and molten metal can cause a fire. Many of the large industrial fires in this country have been caused by careless welding and cutting operations.

Overheating · Damage to electrical conductors can result from heat, exposing them in two ways: the fire that heats the conductor from the outside, and overheating from an internal source. External heating can result from a fire that begins to melt a copper conductor by the constant application of heat. When the wires have reached their melting point (approximately 2,000°F), blistering will occur on the surface of the wire. Shortly thereafter, the wire will begin to distort, usually by elongating or thinning.

Internal heating is most often caused by continuous overcurrent or short circuits. However, the short circuiting is often the result of the continuous internal overheating. As the conductor begins to overheat, it will transmit this heat throughout the circuit. The first result will be the softening of the insulation, which may be rubber or plastic. If plastic, it is usually polyvinyl chloride (PVC). After the insulation is softened, the continued heat will begin to char and crack it, eventually destroying it. As the insulation is ruined, adjacent conductors can contact each other, creating short circuits and the resulting arcing. The bare wire may also make contact with the metal conduit, metal-clad cable, ground wires, water pipes, or even gas pipes.

If, for some reason, the damaged conductor does not short out and the overheating from the excessive current persists, the temperature of the wire may eventually reach its melting point. At this point, the conductor will begin to blister as the heat builds up. When the melting point is reached, the wire will separate. Then the conductor will begin to cool, since the current is no longer flowing through it. However, at the time of separation, there may be an arcing. Any arcing that occurs will be likely to cause movement of the wire. The conductor will probably bead on the separated ends. (The beads on the ends of the wires in Figure 6-9 are examples of this.) As the wires move, they may create additional short circuits, with the attendant arcing. The intense heat generated by the arcing will melt and splatter the copper of the wire.

As current flows through a conductor, it generates heat, in much the same way that water flowing through a fire hose develops friction loss. The smaller the conductor and the greater the amount of current flowing, the more heat generated. This heat is dissipated through the insulation on the conductor. If the insulated wire is exposed to the air and the heat is not excessive, the wire will not become overheated. But if the insulated wires are covered, such as with a rug or carpeting, the heat will be confined and build up. The resultant heat may be sufficient to ignite the covering. If not, the rug or carpeting will retain the heat within the insulation of the wiring. This will cause the insulation to dry out. The results will be the same as previously described—short circuiting, arcing, and a resulting fire.

With properly installed wiring, overheating should not occur. The overload protection by fuses and circuit breakers should function. However, poor or faulty installation can assist in negating the safety devices.

There are several other ways of causing overheating with electrical wiring. Some of the most common are in connection with the misuse of extension cords and unprotected or bare light bulbs. Such bulbs create a problem where grain, sawdust, or any similar combustible materials are in storage. Wherever an unprotected, bare, lighted bulb is permitted to lie on or be buried within these materials, the heat generated by the bulb may be sufficient to cause ignition. The old-fashioned bare bulb hanging from the ceiling in the coal bin was the cause of a number of fires, when the coal bin was filled with the light left on. The lag time or delay on ignition from these conditions is usually from three to six hours.

A fire may also originate from a bare lighted bulb exposed to other types of combustibles. In this case, the glass of the bulb may be heavily stained by the charred fragments or ash of the combustible material. If the bulb was broken during the fire or fire-suppression activities, the investigator must be alert to find any fragments. If the light bulb is intact, it may be easier to locate. The investigator should be aware that light bulbs are not very often accidentally buried in combustible materials.

Electrical appliances such as coffee makers, deep fryers, toasters, hair dryers, and similar modern devices also overheat. They are usually controlled by bimetal thermal or thermostat controls. These should be checked if the appliance is suspected of being the cause of the fire. See if they have become fused in a closed position or are pitted from arcing. If so, these conditions may have permitted overheating. A college student was having trouble with her hair dryer, which was continually shutting off. She took it to an electric-appliance repair shop. The repairman applied a drop of solder on the bimetal strip, effectively overcoming the problem. The student left the dryer on her desk, plugged it in, and went to class. The appliance overheated and started a fire in her dormitory room.

Overheating of electric motors · When an electric motor is suspected as the cause of a fire, check the movement of its shaft. A shaft of an electric motor frozen to the bronze bearing (bushing) is strong evidence of the motor's burning internally. If, for some reason, the steel shaft begins to heat the bearing excessively, the surface temperature of the bearing will build up. Somewhere in the neighborhood of 1,800°F, depending upon the tin content of the bronze bearing, the melting point of the bronze will be reached and the steel shaft can become frozen to the bearing. At these temperatures, all combustibles within the motor housing will have become ignited. Fire may spread to nearby combustibles.

Heat from a fire on the outside of the motor will not usually be sufficient to melt the bronze bushing. Most motors are located in the lower third of the room. Normal temperatures of a fire rolling across the ceiling will be in the area of 1,400° to 1,600°F. But temperatures in the lower third of the room will be only about 500°F, assuming no accelerants are present. The housing of the motor could become very hot, but probably not hot enough to melt the bushing. Always carefully check the fuses and/or circuit breakers servicing any motors suspected of being the fire cause. Be sure they are the proper size and have not been tampered with.

In one case, two motors were suspected as the origin of a fire in a large henhouse (5,000 hens). One motor had a cast-iron housing, the other an aluminum-alloy housing. The latter had been distorted and partially melted during the fire; however, the shaft of its motor turned freely. The motor with the cast-iron housing had a completely frozen shaft, and it was determined that this had been the cause of the fire. The heat generated by the frozen shaft ignited nearby combustibles, which in turn caused the aluminum-alloy housing of the second motor to melt.

A motor that has not frozen will be quite likely to throw bits of molten solder out of the armature or commutator as they turn during a fire. These are usually found on the inside of the housing or on the surface immediately below the motor.

Fluorescent fixtures · Trouble in fluorescent fixtures usually starts with the breakdown of the insulation in the ballast. This causes a short or ground, which generates a great deal of heat and has these results:

1. Molten material contained in the ballast begins to drip. This would be evident on stock, fixtures, desks, equipment, or floors.
2. The pitchblende compound (combustible) develops into a gaseous state and reaches its ignition temperature from the overheating, then explodes. The results are shattered glass and tubes that can seriously injure personnel and damage property.
3. The molten compound or gas reaches its ignition temperature and starts a fire that communicates to any combustible materials on the ceiling or in the vicinity of the fixture.
4. The short continues until the branch-circuit fuse blows. This will cut off all the fixtures and equipment on that circuit.

The odor given off by the breaking down of the ballast is a familiar one to most experienced firefighters. They will recognize it immediately and can advise the investigator that the odor was present. Proper fusing should keep the faulty ballast from causing a fire.

Electrical devices that assist the investigator · Often, the timing of the spread of a fire can be of value to the investigator. Electric clocks, even though damaged or destroyed, can be of assistance. They should be carefully removed from the debris of the fire and closely examined. The time they stopped will often be indicated, and this can help establish the time of the fire spread.

Other electrical equipment that can assist in a similar manner are the steam, fuel, and water graphs that record time, temperature, and flow of materials. Most operate on a 24-hour basis. These recording graphs are often found in a separate room, which may have been unaffected by the fire. In an industrial fire, it has been possible to establish the time and spread of the fire by use of these recording devices.

FLAMMABLE LIQUIDS

Misuse of gasoline and other flammable liquids is one of the major causes of accidental fires in this country. The use of flammable liquids as cleaning agents is still common. Using open containers of gasoline to clean engine parts, machinery parts, paint brushes, and other greasy items inside a structure has resulted in many accidental fires and explosions. The ignition may be either the result of the vapors coming in contact with an ignition source or from a static spark. The most commonly used flammable liquids in this type of incident are gasoline, benzine, petroleum ether, paint thinner, and kerosene.

Accidental spills of gasoline or other flammable liquids have resulted in many fires. Some of these occur from the overfilling of tanks, misfilling of tanks (filling the basement instead of the tank), rear-end collisions of motor vehicles, and collisions of flammable-liquid tankers with other motor vehicles or trains.

Another all too common practice is the storage of gasoline in glass containers in basements, utility rooms, and garages. Many accidental fires and explosions have resulted from the breaking of these containers. In one case, children were skating in the basement of a house; one struck a glass gallon jug of gasoline and broke it. Instead of getting the children out, the mothers grabbed a mop and rags to wipe up the spill. The vapors were ignited by the pilot light of the gas-fired hot-water heater. Four children and two mothers were critically burned and scarred for life. It is the sad duty of the fire investigator to investigate and deal with these incidents.

At construction sites, fuel oil may be used as the fuel for portable heaters and salamanders. When these are left unattended, they can overheat and start fires in the structure.

NATURAL GAS

Accidental fires and explosions from natural gas most generally occur from ignorance, carelessness, gas leaks, and attempts at suicide.

In recent years, a number of gas utility companies have been using plastic pipe to replace old cast-iron, wrought-iron, or steel pipes. This plastic pipe is sometimes inserted in the older and larger mains and is secured to the metal main with compression couplings. During cold weather, the plastic pipe contracts. The National Transportation Safety Board has reported that plastic pipe will contract one inch per 100 feet for each 10°F drop in temperature. The length of plastic pipe involved in two incidents at Fremont, Nebraska (NTSB-PAR-76-6), and Lawrence, Kansas (NTSB-PAR-78-4), was over 300 feet in each case. As a result, in several cases, these plastic pipes have pulled entirely out of the compression couplings, and the result was a wide-open main. Fires and explosions have been the result of these conditions.

Ignorance and carelessness combined to produce misfortune in these two cases: A house was sold and the occupants moved out. They shut off the gas at the meter and removed their gas stove, but failed to cap the gas pipe in the kitchen. The new owners, in order to do some cleaning, needed hot water, so they turned on the gas and lit the gas-fired hot-water heater. While the water was heating, they left the house to do some additional packing. Fortunately, therefore, no one was in the house when the explosion occurred. Figure 6-12 is a photo of the house after the explosion.

FIGURE 6-12. Typical pattern of natural-gas explosion in dwelling. The roof is lifted into the air; the exterior walls fold out, and the roof then drops back on any partition walls that remain standing.

In another incident, the home was sold, but the gas was left turned on. Next to the kitchen stove, whose pilot light was on, was a built-in gas-fired oven. The pilot light of the oven was turned off, but the gas service to it was still on. There was a delay of four days before the new owners could move in. During the forenoon of the third day, the next-door neighbor saw smoke coming from under the eaves of the house. She called the fire department. The firefighters found the kitchen in flames and smoke throughout the house. After the fire was extinguished, the oven was found lying on its face on the kitchen floor. Was it vandalism, arson, or an accidental fire? A fire investigator was called in and found that the fire had been ignited by the pilot light of the stove. The fuel was gas from a pinhole leak in the flexible tubing to the oven. The gas had accumulated behind the oven and eventually reached an ignitable mixture. A low-order explosion pushed the oven over. Figure 6-13 shows a corner of the kitchen with the oven lying face down on the floor. Note the smoke and char pattern on the wall to the left side of the picture.

The wall of the living room directly behind the oven and stove was con-

FIGURE 6-13. Corner of kitchen where built-in oven stood prior to explosion. Oven can be seen lying face down on floor in lower left corner of picture; stove is in lower right. Note charring on wall and corner of wall.

structed of plasterboard. When a dry-wall type of construction is subjected to a low-order explosion from within the partition or from the opposite side of the wall, the wall is pushed out. When the wall is pushed out, so are the nails holding the wallboard to the studs or furring strips. When the wall snaps back, these nails remain in the position to which the explosion pushed them. Figure 6-14 is a photo of the wall on the opposite side of the partition from the kitchen. The nails can be seen extending out of the wallboard. An investigator can immediately recognize these extended nails as an indication of a low-order explosion.

In this particular case, the amount of gas leaking was so slight that it took nearly three days to accumulate a combustible mixture. Previously, the odor of gas had never been noticed, as it was so slight. People moving about the kitchen stirred up the air sufficiently so that it did not accumulate.

FIGURE 6-14. Plasterboard wall on opposite side of partition from kitchen where explosion occurred. Note nail heads extending out from surface of wallboard, indicating that a low-order explosion occurred on the opposite side of the wall or within the partition.

Another case occurred when a man decided to commit suicide because his wife had taken the children and left him. He shut off the gas stove, the gas hot-water heater, and the gas-fired furnace, then opened three union fittings in the gas lines in the basement. The one pilot light he forgot was on the gas-fired clothes dryer. He sat upstairs in the kitchen at the table, writing his farewell letter, as the gas accumulated. Since the clothes dryer was in the basement and natural gas is lighter than air, he waited quite some time before the explosion. There was sufficient gas in the kitchen and living room to cause an explosion that blew one side of the house out. Figure 6-15 is a picture of the remains of the living room and kitchen after the explosion. The man died within 24 hours from burns.

CHEMICALS

Chemicals may be involved in both accidental and incendiary fires. Many types of chemicals are readily available in drugstores, service stations, agricultural supply stores, chemical laboratories, beauty salons, and many other retail outlets.

FIGURE 6-15. Gas explosion resulting from a suicide attempt blew out wall of kitchen and living room. Note high burning and partition wall blown out by gas that accumulated in wall.

Oxidizing agents · Oxidizing agents are chemicals that contain oxygen or combine with other chemicals to give off heat. There are a number of chemicals that may be classified as oxidizing agents, but knowing the names of all of them would require an extensive chemical background. However, certain chemical names, when used as suffixes, indicate a probable oxidizing agent. It would be well for the fire investigator to memorize some of these in order to recognize them readily. Some of the more common names of this type are:

Bromates	Nitrates	Peroxides, inorganic
Chlorates	Nitrites	Peroxides, organic
Chlorites	Perborates	Permanganates
Hypochlorites	Perchlorates	Persulfates

Other oxidizing agents, whose names are not usually found as suffixes, are liquid oxygen (LOX), chlorine, fluorine, and bromine.

When these oxidizing agents come in contact with organic fuels, spontaneous heating or ignition can occur. Some fuels will react with one oxidizing agent, some with another. Some will react with almost any type of oxidizing agent; examples are turpentine, fuel oil, diesel fuel, rubber dust or powder, oily cotton waste, polyethylene powder, paper, sawdust, and pine oil.

One very common oxidizing agent is calcium hypochlorite. This may be found in powder, granular, tablet, or liquid form. It is available at hardware stores and swimming-pool supply stores—in fact, anywhere swimming-pool supplies are available. It is commonly used for chlorinating swimming pools. It is also used by water departments in repairing water mains. The calcium hypochlorite is placed in the repaired or new main, and when the water flows into the main, it releases a great deal of chlorine.

This chemical may be stored in a garage along with an organic fuel, and it takes only an accident to mix them together. Calcium hypochlorite reacts violently with brake fluid and almost all carbonaceous fuels. Two tablespoones of brake fluid poured into a can containing about two cups of calcium hypochlorite crystals will send a sheet of bright red-orange flame four to six feet in the air. The color closely resembles that of a railroad flare. Depending upon the temperature, humidity, and wind conditions, the ignition will take place in from 45 seconds to just under five minutes. After the fire, all that remains are some crusty calcium salts.

Chlorates and perchlorates, which contain both oxygen and chlorine, are usually encountered with sodium or potassium salts. However, sometimes calcium salts, which are used as disinfectants, weed killers, and antibacterial agents, will be encountered. Chlorates spilled on wooden floors have been known to cause fires from the friction of someone walking through the spill.

Potassium permanganate is a common medicinal agent and germicide, containing potassium, manganese, and oxygen, that can be purchased at the drugstore. It is capable of starting a fire when mixed with the organic fuels listed above. The speed of ignition again depends upon temperature, humidity, and wind conditions. The reaction may begin slowly but builds up speed with the exothermic reaction and may result in an explosion.

One indication that oxidizing agents may have been involved, either accidentally or intentionally, is an explosion followed by a fire. Another is a very hot fire observed or evident at the point of origin. If oxidizing agents are suspected, the investigator must look for chemical residues. The salts of potassium, manganese, calcium, and some of the metals, which would not normally be found in the location or occupancy, are possible evidence of the use or misuse of oxidizing agents. The color of the flames early in the fire in the area of origin may indicate the presence of these salts—violet or purple for potassium, reddish-yellow for calcium, greenish-yellow for manganese, and white for magnesium. Sodium burns with a yellow flame, but it is usually covered by the natural flame colors of a fire. The first-arriving firefighters may be able to assist the investigator if they observe these colors.

Nitrates are found in fertilizer and chemical plants. The most common is ammonium nitrate, fertilizer grade (AN-FG), which is readily available as a fertilizer. When ammonium nitrate FG is mixed with an organic material, it can become a blasting agent. Most often it is mixed with fuel oil, in which case it is usually designated as AN-FO. The ingredients of AN-FO are easily obtained and mixed. Many farmers now use AN-FO or commercial blasting agents made from ammonium nitrate in place of dynamite. An explosive booster such as dynamite or a specially prepared commercial booster and a blasting cap must be used for initiation.

The greatest danger connected with ammonium nitrate FG is its being confined and exposed to fire. The Texas City disaster resulted from a fire involving burning ammonium nitrate FG in the hold of a ship. When exposed to fire and heat, it will vigorously support combustion, since it is a strong oxidizing agent.

Ammonium nitrate can become sensitized both by heat and by becoming contaminated with organic materials, such as fuel oil, charcoal, flammable liquids, and such solid materials as ground nutshells. Once this has occurred, it becomes a blasting agent. One of the commercial blasting agent preparations contains ammonium nitrate, fuel oil, and ground-up acorn shells.

In addition to being an oxidizing agent, ammonium nitrate has two other characteristics with which the investigator should be familiar. It is both *hygroscopic* and *deliquescent*. A chemical is hygroscopic when it absorbs moisture from the air and from any substance it touches. People working with it find that it draws perspiration from the body, and their clothing becomes soaked within a short time. As the ammonium nitrate absorbs moisture, part of the chemical will liquefy. This process is known as deliquescence. The liquefied ammonium nitrate will impregnate any nearby combustibles with the oxidizing salt. Consequently, the investigator should be aware that any carbonaceous fuels capable of absorbing the oxidizing salts now have their own oxygen. This would permit the burning of the impregnated materials under conditions that might not otherwise be considered normal.

STEAM

Wood in close proximity to uninsulated or poorly insulated steam pipes will absorb heat from the pipes. If the pressure of the steam is known, the steam-pipe temperature is also known. (See Table 6-1.) For many years, reports of fires caused by the exposure of combustibles to steam pipes have been recorded. The theory has held that the continued application of heat to wood converts it to charcoal, which eventually ignites spontaneously.

The U.S. Forest Products Laboratory reported on experiments conducted with small, kiln-dried maple wedges.[2] Samples exposed to 222°F (107°C) for 1,050 days assumed a light chocolate shade. Those exposed to 248°F (120°C) for 1,235 days were appreciably embrittled and were of a dark chocolate color. Those exposed to 282°F (140°C) for 320 days had the appearance of charcoal.

Their report indicates further that, although ignition did not occur at any time during the experiments, there is no guarantee that it might not take place if the conditions were favorable. Decomposition proceeds very slowly at lower temperatures, and the gaseous products evolved are dissipated. However, in a confined space, where the gaseous products and the heat produced by oxidation could not escape, spontaneous ignition might occur. How long this would take is a matter of conjecture, but the fire investigator must recognize the possibility.

TABLE 6-1

Pressures and Temperatures of Saturated Steam

Gauge Pressure psig	Temperature F°	C°
0	212	100
10	240	115
15	250	122
20	259	126
30	274	135
40	287	142
50	298	148
100	338	170
200	388	198
500	470	243

[2]*Ignition and Charring Temperature of Wood, Report No. 1464*, rev. (Madison, Wis.: U.S. Forest Service, Forest Products Laboratory, 1958), pp. 2–3, Table 2.

Static electricity is not caused by friction in the sense that two materials must be rubbed together. Rather, it is the result of contact and separation of similar and dissimilar materials. As this occurs, one takes a positive charge and the other a negative charge. Industry provides such a static generator in belt drives, unrolling paper, and the movement of materials that contact and separate. Flammable liquids develop static charges as they pass through pipes, hose, nozzles, and even the air.

Most solvents are poor conductors of electricity and consequently allow a charge of high intensity to develop as they move. Water or any liquid, forming droplets and falling through space, can generate a static charge. In nature, this is manifested in the formation of lightning.

Many people are under the impression that humidity stops the generation of static electricity. However, it is during a rainstorm that nature demonstrates static electricity in the form of lightning. Humidity does tend to control static electricity by forming a moist film that condenses on everything in the area. This film contains dust, CO_2, and other materials, all of which are conductive. These in turn make the film sufficiently conductive to bond all surfaces together electrically. This neutralizes the static charges.

Nature can be improved upon by controlling the humidity (usually 60 percent or better) and providing conductors to connect or bond together all machinery and objects in the hazardous area. These precautions will prevent any object from developing a high potential or static charge. Employees in these areas should also be "bonded" to all objects. This is accomplished by using shoes with conductive soles and conductive floors such as those found in hospital surgical suites and delivery rooms. Conductive floors are also common in laboratories and in munitions and explosives manufacturing works. Any place where a combustible gas, dust, or vapor or an explosive atmosphere is likely to develop may have conductive floors.

Improper grounding and bonding have caused a number of fires and explosions. The fire investigator's understanding of the principles of static electricity will greatly assist him in determining if, in fact, a fire or explosion had such a cause.

SPARKS AND EMBERS

The energy crisis of recent years has created the need to provide a more efficient means of heating the home. The use of fireplaces and other wood-burning devices has increased, and so has the number of fires started by these wood-burning devices. The investigator cannot overlook the possibility that sparks and wood embers were thrown out of the fireplace, particularly when no screen was used. Poorly constructed hearths, fireplaces, and chimneys are all likely places for sparks or embers to find combustible materials. Old chimneys and fireplaces are notorious for cracks and other openings into hidden areas such as wall interiors and attics.

Some manufactured fireplace logs are not recommended for the metal types of fireplaces and stoves. These logs burn with much greater intensity than wood does and will overheat this kind of equipment very quickly. Fires have resulted from their misuse.

The accumulation of soot and similar products of combustion requires periodic cleaning of chimneys, both masonry and metal. Failure to clean these properly can result in a chimney fire, as can inadequate clearances.

In addition, the returning popularity of wood-shingle roofs and the mansard type of cornice covered with cedar shakes has created additional fire hazards. As these combustible roofs and exterior attachments grow older, their susceptibility to exposure fires from sparks and embers increases. Many contain concealed spaces that permit hidden communication of fire. An increase in the number of roof fires may result.

A roof fire requires the investigator to determine whether it burned from the outside in or from the inside out. Fires communicating from within will usually produce much greater fire damage over a larger area inside the structure. Once the interior fire has breached the roof, most heat and fire will be drawn out of the hole. The intensity of the fire will increase. Any portions of the roof not destroyed will show the effects of the fire on the underside. The exterior roof fire burns differently. It will spread over the surface of the roof, and if it breaches, the burning around the hole will be gradual rather than the extensive damage from the interior fire.

Another type of spark that causes fires is that from a diesel locomotive; it can ignite combustible materials along railroad tracks. During dry periods, these fires can create very serious problems.

Grinding and sanding operations also produce an abundance of sparks. Some of these operations can throw sparks as far as 50 feet. These activities are often conducted in atmospheres containing combustible dusts or flammable vapors. Drilling and machine-press operations may also throw off pieces of hot metal.

Rubbish burning in or outside incinerators can produce sparks and embers. Several large-loss fires have started from burning rubbish blowing under combustible loading docks. One of these, in Minneapolis in 1955, started 70 additional exposure fires. Also produced from rubbish fires are glowing fragments of wood, paper, and cardboard, which may travel considerable distances. During the Minneapolis fire, the wind blew embers that started fires a mile away from the original fire.

Small fragments of burning tobacco from cigarettes, cigars, and pipes also have started fires. Furniture and automobile seats are particularly susceptible to this type of incident.

MATCHES

Accidental fires from matches occur through carelessness in striking the match, failure to close the cover of the matchbook (the entire matchbook can burst into flames), burning particles from the head of the match flying off and hitting combustible materials, or the entire burning head of the match separating from the stick and hitting combustibles. The large, strike-anywhere matches have ignited in pockets when the match heads were hit together or struck some abrasive material in the pocket.

COOKING AND HEATING

As everyone working in fire prevention is aware, clearances play an important part in preventing fires. Most restaurant fires originate in the area of the deep fryers, grills, ranges, and other equipment that are under hoods, or wherever considerable greasy vapors are present and can accumulate. Permanently installed automatic fire-extinguishing systems in hoods and vents have reduced serious results in many of these fires. But hoods, ducts, vents, and filters must be kept clean. When these are full of grease, fires will occur. And 24-hour restaurants seems never to have the time to shut down and clean things up.

Many of these hood and duct fires are undetected until the extinguishing system operates or the fire has spread. Without an extinguishing system, fires will communicate up the duct. Sometimes the ducts extend to the roof of a high-rise building, with fire all the way up.

Large and small city fire departments have many calls for "meat [burning] on the stove." Unattended cooking has been the primary cause of quite a few fatal fires. And in some of these fires, the investigator must expend a great deal of effort and energy.

There are still many of the old pot-type oil burners in use. These must be cleaned periodically. In one case, the oily soot deposit in the pot was about two inches thick when the unit blew up. Oil-fired and gas-fired furnaces have several safety devices, each of which must be checked out. If there is any question in the mind of the investigator, he or she should bring in experts to assist in the investigation.

Sometimes heating equipment is missing the label. Always look for the UL label (in the case of oil-fired heating units) or the AGA label (in the case of gas-fired units). If the label is missing and the heating unit is suspected as being the cause of the fire, the investigator should try to find out why the label is missing.

OTHER ACCIDENTAL FIRE CAUSES

Although the use of candles about the home has diminished in the past 25 years, it is now becoming more common. For many years, the investigator was able to detect the wax drippings from candles. Today, with the dripless candle in common use, they are not as easy to detect. Church candles, party candles, and those used in the home for "effect" have all contributed to fire losses. One type of party candle is the non-extinguishable birthday cake candle. No matter how hard one blows, it will not go out. The wick is impregnated with chemicals that prevent extinguishment by blowing on it.

The thawing of frozen pipes with a blow torch can start a fire from overheating adjacent combustible materials to their ignition temperature.

A few years ago, it became very popular to line family rooms, recreation rooms, and hallways with redwood, pithy cypress, and various plywoods and veneers. To maintain the effect, these walls were waxed and polished. Fire, feeding on the wax and polish as well as the woods, spreads rapidly through these rooms and hallways.

Some plywoods are constructed with adhesives that react to relatively low temperatures. The heat causes the adhesive to release the plies or veneers of wood, which then delaminate. As they are extremely thin, they will readily ignite. Fire will flash across the room or down the hall. Quite a few fatalities have resulted from fires of this type.

Incendiary Fire Causes and Effects

When all accidental and providential causes have been eliminated, the investigator begins to look for an incendiary cause. This requires the answers to a series of questions about the fire. Again, look for the unusual, the absence of the normal, the presence of something different from what one would normally expect for the occupancy. Some of the things that will fall into these categories were listed in Chapter 5. In addition, are all pets accounted for? In mercantile occupancies, has there been a substitution of junk or secondhand merchandise for the type that would normally be expected?

PLANTS AND TRAILERS

The investigator should be looking for plants and trailers. The word *plants* refers to the preparation and gathering of material to start a fire. Newspapers, combustible plastics, rags, clothing, curtains, blankets, cotton waste, wood shavings, and other combustible materials have been used as plants.

An accelerant is often used in connection with the plant. The most commonly used accelerants are gasoline, fuel oil, paint thinner, and lighter fluid. These may be sprayed or poured on the piled combustible materials. A plant may be placed around an ignition device, which may range from something as simple as matches or a candle to some sophisticated electronic mechanisms. The most common ignition device is the match. When searching the point of origin, do not overlook unburned or partially burned matches or matchbooks. Any of these found at the point of origin should be carefully preserved for examination by the laboratory. Unburned or partially burned matches left at the scene of a crime have been identified as coming from a partial pack of matches found on the suspect. Sometimes, after thoroughly sifting the ashes and debris at the point of origin, the investigator will find only a staple remaining from a book of matches. This too is evidence.

In order to spread the fire within the room or throughout the structure, a device called a *trailer* is used. Materials used for trailers are limited only by the imagination of the firesetter. Flammable liquids are frequently used for this purpose, by pouring them on the floor or floor covering. Cotton rope soaked in kerosene or gasoline, newspapers, toilet paper, facial tissues, paper towels, combustible packing materials, almost anything that will support combustion once ignited may be used. These trailers will lead from one plant to another, to furniture, even to holes in walls.

The investigator must be able to recognize plants and trailers. Some are very obvious, others well concealed in their makeup.

FIGURE 6-16. Fire was started with "plant" of flammable liquids and combustible materials alongside of cardboard cartons of vigil candles in church.

On some occasions, the firesetter will knock holes in the walls or tear open the ceiling to assist the spread of fire. The firefighter may knock similar holes in walls and pull ceilings in a like manner during fire-suppression and overhaul operations. It will be necessary for the investigator to determine if holes in the walls or ceiling were made before or after the fire started. If trailers or charring from trailers leads to these holes, the investigator's job may be easier. Figures 6-16 and 6-17 are illustrations of trailers and plants.

Another condition, which can be misinterpreted, is holes burned in the wood-lath and plaster wall. These may occur as heavily charred, isolated spots in the wall itself. They may be holes poked in the wall prior to the fire to assist in its spread. Or they could be the result of the "scratch" coat of the plaster deteriorating and falling out of the wall. The "scratch" coat is the base or first coat of plaster applied to the wood lath. As it dries out, pieces of it may occasionally drop out from between the laths. The cavity formed in the wall may

FIGURE 6-17. Flammable-liquid trailers on kitchen floor are clearly visible. Note the pattern of the charring on the floor. (Racine Fire Department photo, by Lt. William S. Jones, Jr.)

not be visible if the wall is covered. But as the fire develops in the room or within the wall, combustible gases are generated. These gases fill up the cavity and, when ignited, burn a hole in the wood lath at that point. Figure 6-18 illustrates this type of spot fire.

The inspection manual of the International Fire Service Training Association says:

> The thickness of interior wall coverings, plywood, dry wall, and plaster may be measured and if the fire has penetrated the wall, an approximate time factor can be established from known fire-rating thickness for a fire of average temperatures. An examination of the interior of a wall can be a very confusing experience if the wall finish has been entirely removed. In some instances, a variety of types of gypsum board is used in combination with paneling, painting, and papering. Each of these materials will provide a different degree of protection to the studs and structural members.
>
> Some of the more common types of interior wall coverings with their corresponding thicknesses and fire resistive ratings are shown in the follow-

FIGURE 6-18. Spot fires burned in old wood-lath and plaster wall.

ing table. The factors given are derived from testing under controlled laboratory conditions and may only be used as a guide because of the many
variable factors involved in the average structure fire.[3]

Wall Covering	Thickness	Resistive Rating
Fir Plywood	$\frac{1}{4}''$	10 min.
Fir Plywood	$\frac{3}{8}''$	15 min.
Fir Plywood	$\frac{1}{2}''$	20 min.
Fir Plywood	$\frac{5}{8}''$	25 min.
Gypsum Wallboard	$\frac{3}{8}''$	25 min.
Gypsum Wallboard	$\frac{1}{2}''$	40 min.
Gypsum (Two layers)	$\frac{3}{8}''$	1 Hr.
Gypsum (Two layers)	$\frac{1}{2}''$	1 Hr. 30 min.
Plaster on Wood Lath	$\frac{1}{2}''$	30 min.
Plaster on $\frac{1}{2}''$ Perf. Gypsum Lath	$\frac{1}{2}''$	1 Hr.

Another method the arsonist uses to create a good draft for the fire spread
is to open or break windows or skylights; skylights are not as obvious as windows.
Again, it will be necessary for the investigator to determine if the windows or
skylights were broken prior to the fire or by fire-department operations. If the
fire has heavily stained the windows, broken unstained glass may be helpful in
making this determination. The window shown in Figures 5-15 and 5-16 is a
good example of this type of condition.

WINDOW-GLASS CONDITIONS

The IFSTA inspection manual states:

Heavy baked-on smoke stains and a varnish-like or flat black paint appearance with large cracks in the glass indicate a long-smoldering fire with a
proportionately slow heat buildup as illustrated in [Figure 6-19]. Heavy
baked-on smoke stains with varnish-like flat black paint appearance, combined with crazed glass, is indicative of a fire which was initially a slow,
smoldering fire that may have vented itself or otherwise accelerated and
resulted in a fast or rapid heat buildup during its latter stages. The crazed
condition noted on window glass is the result of heat-caused stresses acting
against the inherent tension of the glass. The exterior surface of the glass,
being a relatively constant temperature and not expanding as rapidly as the
interior surface exposed to heat, builds up severe stress and the glass bursts

[3]*Fire Prevention and Inspection Practices*, 4th Ed. (Stillwater: International Fire
Service Training Association, Oklahoma State University, 1974), p. 74.

FIGURE 6-19. Heavy baked-on smoke stains on glass of door with large cracks in the glass, indicating the slow buildup from a long-smoldering fire.

into a crazed state when suddenly cooled by a stream of water spray. These crazed check marks can be seen on the window in [Figure 6-20]. Heavy carbon or soot deposits which wipe off easily combined with large cracks is a condition which indicates a short- or medium-duration fire, possibly accelerated with a hydrocarbon such as gasoline.[4]

METALS AS TEMPERATURE INDICATORS

Many times, it is important that the fire investigator get some idea of the temperatures produced during the fire. If the heat was greater than would normally be expected for the fire loading of the occupancy, the investigator will want to be able to answer the question, Why?

The melting points of various metals may assist in determining what range of temperature was developed. One of the more common metals found in many

[4]*Ibid.*, pp. 71–72.

FIGURE 6-20. Crazed check marks resulting from droplets of water hitting hot glass.

occupancies is copper. It is used in wiring and plumbing. Its fairly high melting point of 1,981°F (1,083°C) means that it will require a very hot fire to cause it to melt. This temperature is much greater than that normally found in a fire of the dwelling type of occupancy. Aluminum, another common metal in the home, melts at approximately 1,220°F (660°C). This is below the temperatures normally found at the ceiling and upper areas of the room. However, if melted aluminum was found in the lower portions of the room, the investigator may want to know how over 1,200°F of heat developed at that point.

FLAMMABLE LIQUIDS AS ACCELERANTS

Flammable liquids are frequently used as accelerants in incendiary fires. Gasoline, fuel oil, paint thinner, and lighter fuel are most common, but any flammable liquid may be used. Some do not lend themselves to this use because of volatility or high flash point. Those with high flash point are difficult to ignite.

Many firesetters have used gasoline but, not understanding its volatility, have been seriously burned when they ignited it. If an excessive amount of gasoline has been used as an accelerant, the investigator might consider a check of hospitals, medical clinics, and doctors for a burned firesetter, or beauty salons or barbershops where someone with burned hair may have sought assistance.

In any building, few floors are perfectly level. Flammable liquids will tend to puddle not only in doorways, but in any low spot, such as the corners of the room or along the wall on one side. The dead air space of the corner of a room is one of the least likely places to be damaged when no accelerants have been used. Therefore, severe burning in a corner of a room or along a wall can indicate the possibility of an accelerant. An exception to this would be a fire that communicated from below by moving up the wall or in the channels between the studs adjacent to a corner. Figure 6-21 shows the results of a fire that started in the basement from an overheated furnace. This fire burned out the first-floor joists from below and dropped the rest of the floor into the basement. Note that most of the floor was destroyed by the fire from below. In spite of this, a portion of the floor in the corner and along one wall remained intact and unburned.

FIGURE 6-21. Corner of room left unburned in fire that started from overheated furnace in basement. Most of the floor was burned or dropped into the basement.

Another characteristic of flammable-liquid use is the burning along the wall down to the floor level and under the edge of the molding. It may burn patterns on the floor, as seen in Figure 5-4; however, Class A combustibles tend to burn above the floor level, as they rarely burn up completely without an accelerant. The ash and unburned material insulate and protect a portion of the floor.

Combustible materials used as draperies, curtains, or accoustical tile may fall down and burn patterns that resemble those of a flammable liquid. However, a careful examination and reconstruction of the area at the location of the burn pattern will help determine what was involved. Figure 6-22 illustrates this point. A fire was set with an accelerant on a couch and on the floor in front of the couch. This fire ignited the drapes behind the couch, which in turn ignited the combustible tile ceiling. Figure 6-23 shows the areas in front of and behind where the couch was located. The pattern on the left was from the accelerant; that on the right, behind the couch, from the drapes.

FIGURE 6-22. Fire started in sofa in front of window drapes, communicated to drapes, then to combustible tile ceiling and to bookcase alongside windows. Note char pattern on wall from burning drapes and sofa.

FIGURE 6-23. Two patterns appear in this picture. The one on the left was from the accelerant used on the sofa and floor in front of the sofa. The pattern on the right was caused by the burning drapes.

Sometimes the firesetter will use more exotic fuels. In the case of a fire in a small church, the accelerant was suspected of being JP-4 jet fuel, because of the extreme heat generated in a very brief time plus the availability of the fuel. The minister had locked up and left the church about 2:30 P.M. The alarm was called in to the fire department about 20 minutes later. When the first fire-department units arrived, the flames were shooting out the back windows of the church. The first line was taken through the open back door into the kitchen, and the fire was rapidly knocked down. However, extreme heat from the accelerant had actually destroyed the porcelain on the cast-iron kitchen sink, as Figure 6-24 shows. The handles on the faucets were melted off by the heat, and the washers inside the faucets were destroyed. The water was running from the faucets at full force when the fire was extinguished. Apparently, some quantity of the accelerant was poured into the sink with the drain blocked. The water was turned on to cause the floating, burning flammable liquid to overflow when the sink was filled. Figure 6-25 shows the remains of the wooden cabinet built around the bottom of the sink. This was apparently destroyed by the flammable liquid overflowing from the sink. Note the severe charring and heavy alligator marks on the studs and legs of the sink.

Figure 6-26 shows the remains of the partition wall between the kitchen and the church office. The plywood wall affixed to the studs was completely destroyed. The only evidence of it remaining was the nails sticking out of the studs. Figure 6-27 shows the severely burned, exposed 2-by-10-inch floor joists of the choir loft, which were above and in front of the sink. It should be remembered that this destruction occurred in less than 30 minutes.

Sometimes an arsonist will attempt to disguise the odor of the flammable

FIGURE 6-24. Old porcelain-covered cast-iron kitchen sink damaged by accelerant. Note the severe destruction of the porcelain in the sink. Studs in back of sink indicate a very hot, fast-burning fire. The severe destruction to the sink could be caused only by such a fire.

FIGURE 6-25. Another view of the sink, and damage done to the wooden cabinet below.

FIGURE 6-26. Remains of partition between kitchen and church office. Note complete destruction of plywood partition.

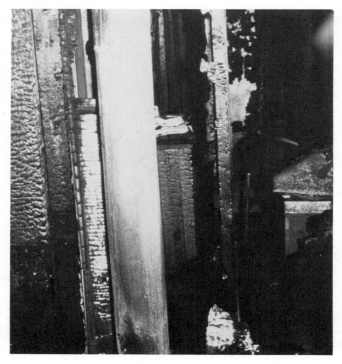

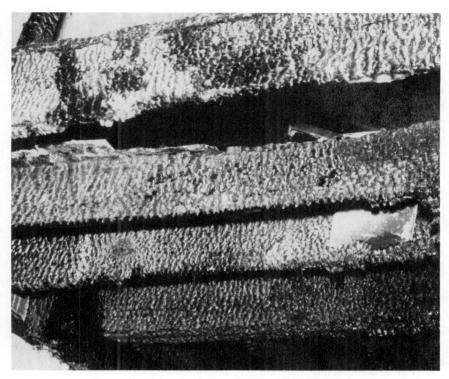

FIGURE 6-27. Severe destruction to 2-by-10-inch floor joists just above and in front of sink.

liquid. This is done by spraying strong odorants, such as ammonia, perfume, or deodorants, in the area to overpower the odor of the accelerant. If firefighters encounter the odor of ammonia when no ammonia should be present, an explanation is necessary.

SUSPICIOUS FIRE ORIGINS AND CONDITIONS

There are a number of areas within a building where the origin of a fire should be considered suspicious. Among these are:

1. Closets
2. Stairwells, or hallways behind or beneath stairs
3. A central area of the building
4. Water heaters
5. Bathrooms
6. Unfinished basement areas or crawl spaces

QUESTIONS

1. Define: *plants, hygroscopic, deliquescent, trailers.*

2. Which of the following are subject to spontaneous heating?
   Animal oil
   Vaseline
   Mineral oil
   Vegetable oil
   Coal

3. When an electrical conductor breaks and arcs, what is the usual result on the broken ends of the conductor?

4. What is the significance of melted copper pipe in a house fire?

5. What is the difference between a "hot" and a "cold" lightning bolt?

6. What is the significance of a frozen steel shaft on a motor in the area of the origin of a fire?

7. What is an oxidizing agent?

8. What are the three sources of most electrical fires?

9. What is indicated by lips on the edge of the hole burned in a conduit by arcing.

10. List six oxidizing agents.

11. What effect can cold weather have on circuit breakers mounted on the exterior wall of a garage?

12. What oxidizing chemicals do chlorates and perchlorates contain?

13. What two odorants are most commonly used to disguise the odor of a flammable liquid?

14. List four areas within a building where the origin of a fire would lead one to be suspicious.

7 Explosions and Their Characteristics

Basic Types of Explosions

There are three types of explosions: mechanical, chemical, and nuclear. Many different names and terms are used to describe them. However, those working in the field of explosives and explosion investigations have agreed on the following standard group of terms:

MECHANICAL

A mechanical explosion is any explosion that can occur only within a container or vessel. This type of explosion must involve an unstable physical condition, which consists of atmospheric pressure on one side of the container and a pressure that is higher or lower than atmospheric on the other side. The condition may actually exist for only a millisecond, as in the case of a pipe bomb. Examples of mechanical explosions are the B.L.E.V.E. (boiling liquid expanding vapor explosion), a pipe bomb, a boiler explosion, an aerosol can, the bursting of a fire hose, a backdraft, or the bursting or popping of a kernel of corn.

CHEMICAL

A chemical explosion is caused by the extremely rapid conversion of a chemical compound into gases. The chemical compound may be either solid or liquid. Chemical explosions occur when unstable chemical compounds (chemical explosives) return to more stable compounds or conditions with great speed. The entire conversion process takes place in an extremely short space of time. It is accompanied by shock waves, a loud noise, and extremely high temperatures.

The chemical explosion can occur either when confined or without being confined. Examples of chemical explosives that result in chemical explosions when detonated are the various military explosives, dynamite, blasting agents such as ammonium nitrate–fuel oil (AN-FO).

NUCLEAR

A nuclear explosion occurs within the atom of an element and may be either nuclear fission or nuclear fusion. The investigation of nuclear explosions requires very special equipment and clothing. Because this type of explosion is highly destructive and damages a widespread area, its investigation is handled by specialists trained in this field.

Basic Results of Explosions

From these definitions, it may be perceived that three almost simultaneous events occur when an explosion takes place:

1. A sudden release of energy, accompanied by light, heat, and noise.
2. A sudden or sharp rise in pressure. This will result in the generation of two pressure waves—one positive, the other negative.
3. The moving of materials.

To clarify, an explosion is the result of an unstable compound or condition returning to a more stable condition with great speed. It will be accompanied by the release of energy, heat, light, and noise. Two pressure waves result from an explosion. The first is a positive wave, which is the force of the explosion traveling away from the center of the explosion in all directions. The second or negative pressure results from the first and is the air rushing back toward the center of the explosion to fill the vacuum created by the passage of the positive wave. The negative wave has about 60 percent of the power developed by the positive wave.

Classification of Explosions and Explosives

During the discussion of explosions and explosives, the terms *low-order* and *high-order* have been used. It is necessary that these terms be clarified so they may be used correctly. They should be used *only* to describe the manner in which the explosives have performed. The term *high-order explosion* means that the explosive being discussed, regardless of type, was totally used up by the explosion. The term *low-order explosion* indicates that, owing to some defect either in the explosive or in the manner in which it was assembled, the entire charge of explosive failed to be consumed and some portion of it remains.

The terms *high explosive* and *low explosive*, omitting the word *order*, refer to the velocity generated by the explosive when it is properly and completely used up in an explosion. (As an example, dynamite is classified as a high explosive. But should a charge of dynamite fail to be totally consumed, it has sustained a low-order explosion.) These terms, in other words, describe the speed of the chemical reaction. In the low explosive, it is called the speed of *deflagration* (burning). In the high explosive, it is called the speed of *detonation*. The result of a deflagration is a "pushing" effect; a detonation has a shattering effect. Black-powder safety fuze burns at the rate of 40 seconds per foot. Black powder is a low explosive. The detonating cord, sometimes called primacord, containing PETN (pentaerythritol tetranitrate), detonates at the rate of 21,000 feet per second. Primacord may be used as an explosive booster or, when wrapped around an object, as a high explosive. PETN and primacord are both high explosives.

An examination of the basic types of chemical and physical phenomena reveals that two may be classified as explosions. These are the low-order and the high-order explosions. Another, depending upon its environment, may or may not be so classified; this is the flash fire, which is a flashover vapor explosion. This occurs when a fire creates an implosion. The implosion in turn creates a vacuum, drawing air into the fire. The additional oxygen causes the fire to flare up, creating the flash-fire phenomenon. In many cases, this could be classified as an explosion, if its volume and pressure were confined. The same fire in the open would be an extremely rapidly spreading fire, pulling in air by the vacuum it creates.

An explosion involving a low explosive is a deflagration. This type of explosion is characterized by a relatively slow "push." A deflagration has a maximum velocity of 5,000 feet per second. Black powder, smokeless powder, dust, combustible mists, and flammable gases and vapors are examples of materials that can explode as deflagrants.

A high-order explosion or detonation is usually:

> ... regarded as an extremely rapid and violent explosion, with a practically instantaneous release of chemical energy. This particular phenomenon has specific characteristics in that it is associated with a shock wave (traveling faster than the speed of sound) and is characterized by a most powerful and sudden disruptive effect, generally evidenced by high fragmentation or great shattering (brisance) and frequently a deep cratering action.[1]

Brisance is determined by the velocity of the detonation wave of the explosion. An explosion involving high explosives can have velocities ranging from 5,000 feet per second upwards. Dynamite, sensitized ammonium nitrate (AN-FO), various blasting agents, and such military explosives as trinitrotoluene

---

[1] Arthur Spiegelman, "Autopsy of an Explosion," *Fire Engineering*, 120, No. 10 (October 1967), 52–53.

(TNT), RDX, C4, MHX, and PETN are all examples of materials classified as high explosives.

COMBUSTIBLE VAPORS

When the ideal mixture of combustible vapor and air is present, ignition will result in a high-order explosion. Under these conditions, there will be a most rapid and complete combustion of the vapors. A low-order explosion may occur from the ignition of a combustible mixture of vapor and air that is close to either end of the flammable (explosive) range—that is, being either too rich or too lean to burn. This results in the combustible mixture not being totally consumed. As in the case of a chemical explosive, this would be considered a low-order explosion.

CHEMICAL EXPLOSIVES

In connection with chemical explosives, a low-order detonation or explosion (an incomplete detonation or complete detonation at below-maximum velocity) may be caused by any one or a combination of several conditions:

1. An initiator of inadequate power
2. Deterioration of the explosive
3. Poor contact between the initiator and the explosive
4. Lack of continuity in the explosive (air pockets within the explosive itself)

Spiegelman lists as some of the more common causes of explosions:

1. *Gases:* Natural gas, sewer gas, other flammable gases
2. *Flammable liquids:* Gasoline, solvents, cleaning fluids, and other low-flash-point flammable liquids
3. *Dusts:* Combustible metal, agricultural material, plastic, and carbonaceous dusts
4. *Explosives and blasting agents* of a commercial variety; also, bombs and incendiary devices
5. *Unstable or explosive chemicals*
6. *Steam, air, mechanical or electrical explosions*[2]

Flammability Limits

A low-order explosion may be a familiar characteristic of natural- and other heating-gas disasters. The vapor–air mixture resulting in a low-order explosion will be found at the extreme lower or upper ends of the flammable limits. Until

[2]Spiegelman, "Autopsy of an Explosion," p. 53.

the percentage of gas in the gas–air mixture reaches the lower limit of flamma-bility—which varies according to the gas, but is about 4 to 5 percent with natural gas (methane)—nothing can happen because the mixture is too lean to support combustion. The upper flammability is the point at which the gas–air mixture becomes too rich to burn (about 15 percent for natural gas). However, between the lower and upper limits, the entire gas-filled area lies ready to explode upon exposure to even a single spark or flame.

Depending upon the quantity of combustible vapor present, in an open space such ignition would usually result in a harmless puff or boom. When con-fined, the almost instantaneous increase in temperature is from ambient to 2,500°F, and upward to nearly 4,000°F. The increase in temperature results in an equally sudden pressure increase. This latter will range from 20 pounds per square-inch gauge (psig), with the lean or very rich gas–air mixture, to over 100 psig for perfect mixtures. Strangely, these pressures do not vary too much between the more commonly encountered fuel gases. Johnson, in discussing fuel gases, says:

> They all generate a pressure of about 20 to 25 pounds per square inch at their lower limits of flammability; rise sharply to maximums of 100 to 115 psig at points nearly midway between their lower and upper limits; and then drop off equally sharply to about the same 20 to 25 psig at their upper limits.[3]

Effects of Explosions

COMMON FUEL GASES AND VAPORS

The effect of an explosion would be about the same with all common fuel gases, either used individually or mixed. It is quite likely that there would be a considerable increase in violence if any of the gases or mixtures were to become ignited near the middle of the flammability range, as opposed to near the lower or upper limit. Few structures are built to withstand even the minimum 2,880 pounds per square foot (psf) without serious damage, if not complete destruc-tion. The 2,880 psf occurs at the extreme ends of the flammability limits. For example, a force of 1 psig equals 144 psf, 2 psig equals 288 psf, and 20 psig becomes, 2,880 psf.

If confined, the pressure resulting from thermal expansion alone may be so great that its force is almost as violent as though an explosion had taken place. However, this is a condition of the particular occurrence and should not be considered to be the inevitable result.

[3] Allen J. Johnson, "Fingerprints of Fire—Investigating Explosive Fire Causes," *Fuels*, Fuels Utilization & Evaluation Letters & Service (Lansdowne, Pa.: By the author, 1958), p. 27.

Thus, the destruction of exterior walls without disturbance of partitions merely shows that explosive gases on both sides of the partitions simultaneously created balancing pressures. By the same deduction, it is possible to determine the exact area that had contained the explosive gases merely by examination of the ruins.

A second frequently encountered example of the mechanical-pressure explosion is the backdraft. The backdraft is the rapid combustion of flammable gases, carbon particles, and tar balloons that have been heated well above their ignition point. This condition develops but does not become ignited because of insufficient oxygen (air). It is most likely to occur during the smoldering phase of a fire. With no flames present, the highly flammable products of incomplete combustion accumulate, and at the same time, their temperatures and the pressure within the room increase. When air or oxygen is introduced, the explosion occurs.

CHEMICAL EXPLOSIVES

Upon detonation, the chemical explosive material is almost instantaneously converted from a solid or liquid into a rapidly expanding mass of gases. The detonation of the explosive will produce three primary effects that can create great damage in the area surrounding the explosion. These effects are blast pressure, fragmentation, and incendiary or thermal damage.

Blast-pressure effect · When an explosive charge is detonated, very hot, expanding gases are formed within approximately 1/10,000 second. These gases exert pressures of about 700 tons per square inch on the atmosphere surrounding the point of detonation, at velocities of up to 13,000 miles per hour, compressing the surrounding air. (Compare these with the explosive forces generated by common fuel gases and vapors.) The mass of expanding gas rolls outward in a circular pattern from the point of detonation like a giant wave, weighing tons, smashing and shattering anything in its path. The further the pressure wave travels from the point of detonation, the less power it possesses, until, at a great distance from its creation, it diminishes to nothing.

This wave of pressure is usually called the blast-pressure wave. The blast-pressure wave has two distinct characteristics; that is, two different types of pressure will be generated: the positive pressure wave and the negative or vacuum pressure wave.

When the blast-pressure wave is formed at the instant of detonation, the pressures actually compress the surrounding atmosphere. The compressed layer of air is known as the shock front. The shock front, only a fraction of an inch thick, is that part of the atmosphere that is being compressed before it is set in motion to become part of the positive pressure wave.

As the shock front, followed by the positive pressure wave, moves outward, it applies a sudden, shattering, hammering blow to any object in its path. Thus, if it strikes an object such as a brick wall, the shock front will deliver a massive blow to the wall followed instantly by the strong winds of the positive pressure wave itself. The shock front shatters the wall, and the positive pressure wave gives it a cyclone-like, sudden, and violent push, which may cause all or part of the wall to topple in a direction away from the point of detonation. The positive pressure wave lasts only a fraction of a second. After striking the wall, it continues to move outward until its power is expended in the distance traveled.

At the instant of detonation when the positive pressure wave is formed, it begins to push the surrounding air away from the point of detonation. This outward compressing and pushing of air forms a partial vacuum at the point of detonation, so that when the pressure wave finally diminishes to nothing, a broad partial vacuum exists in the area. This causes the compressed and displaced atmosphere to reverse its movement and to rush inward to fill the void. This reaction of the partial vacuum and the reverse movement of the air is known as the negative pressure or vacuum wave.

The displaced air rushing back toward the point of detonation has mass and power, and although it is not moving nearly as fast inward as the pressure wave was moving outward, it still has great velocity. This inward rush of displaced air will strike and move objects in its path. It can suck windows out of buildings, and has done so on occasion. Its direction and the direction of the objects it moves are, however, toward the point of detonation.

The negative pressure wave is less powerful than, but lasts about three times as long as, the positive wave. The entire blast-pressure wave, because of its two distinct actions, actually delivers a one–two punch to any object in its path. Its effect is the most powerful and destructive of any produced by the detonation of high explosives.

Fragmentation effect · When an encased explosive such as a pipe bomb detonates, the rapidly expanding gases produced by the explosion cause the casing to rupture and break into fragments. Approximately half the total energy released by the explosion is expended in rupturing the case and propelling the broken pieces of it outward. Fragments resulting from the detonation of a high-explosive filler have a stretched, torn, and thinned configuration, owing to the tremendous heat and pressure produced by the explosion. In contrast, the detonation of a pipe bomb containing black powder, a low explosive, would produce fragments larger in size than those resulting from a high-explosive detonation, and they would not have a stretched and thinned configuration.

Fragments are what we call these pieces of the bomb casing that are formed when it ruptures. Precut or preformed objects such as nails, ball bearings, or fence staples, which are placed either inside the bomb or attached on the outside, are referred to as *shrapnel*. Shrapnel serves the same purpose and has the same effect on people, objects, and structures as fragmentation. One advantage of using shrapnel is that part of the energy released by the explosion, which would normally have been expended in fracturing the bomb casing into fragments, is used instead in propelling the preformed, separate pieces of shrapnel. Consequently, the use of shrapnel inside a bomb or attached to its outside results in an increase in blast damage as well as the projection of the shrapnel. Fragmentation and shrapnel produce damage by cutting, slicing, or punching holes in anything in the vicinity of the point of detonation.

The heat of fragments produced by the detonation of a high-explosive bomb may also cause secondary fires. This heat is induced at the instant of detonation and compounded by the stretching and tearing action of the detonation as well as by air friction and impact friction. The hot fragments may, for example, puncture an automobile fuel tank and ignite the gasoline, imbed themselves in combustible materials and cause ignition, or start grass fires some distance from the point of detonation.

FIRE AND THERMAL DAMAGE

Another point arising from the difference between the lower and upper limits of flammability is their relationship to the amount of fire damage. At the upper flammable limits, approximately three times the amount of heat is released, owing to the greater quantity of gas present. Also, the products formed by the explosion of the gas–air mixture at the upper limits contain large amounts of carbon monoxide and hydrogen. These gases can be almost instantly reignited by any small flames from the initial explosion and create secondary fires and explosions of great intensity. At the other extreme, the products formed consist chiefly of carbon dioxide, water vapors, and unused air, all incapable of combustion. It should be remembered that these same products are the first to be distilled when pyrolytic decomposition occurs. The speed of flame propagation developed by a gas explosion will have a direct bearing upon the extent and the type of damage.

Thermal effect of chemical explosives · The incendiary thermal effect produced by the detonation of a high or a low explosive varies greatly from one explosive to another. In general, a low explosive will produce a longer incendiary thermal effect than will a high explosive. A high explosive will, on the other hand, produce much higher temperatures. In either case, the duration of the effect is measured in fractions of a second. If a high-explosive charge is placed on a section of earth covered by dry grass and detonated, only a vacant patch of scorched earth will remain. However, if a low-explosive charge is placed on the same type of earth and detonated, more than likely a grass fire will result.

Unless highly combustible materials are involved, the thermal effect plays an insignificant part in an explosion. Should combustible materials be present and a fire started, the debris resulting from the explosion may provide additional fuel and contribute to spreading the fire. When fires are started inside a structure that has been bombed, they usually result from broken or shorted electrical circuits and ruptured fuel-gas lines rather than from incendiary thermal effects. The incendiary thermal effects are generally the least damaging of the three primary detonation effects of chemical explosives.

Effects of Vapor Density

As we have seen, one of the properties of a flammable vapor or gas that is of great importance to the fire investigator is the flammable (or explosive) limits. Another property that must be understood is the vapor density of the vapor or gas. This is the relation of a given vapor or gas to air. With air designated as 1.000, any vapor or gas lighter than air will have a vapor density less than 1.000. Likewise, any vapor or gas heavier than air will have a vapor density greater than 1.000. This helps indicate the probable level in an area or structure where the specific gas or vapor might be found.

As an example, because of the difference between the vapor density of natural gas (methane), 0.6, and that of propane, 1.6, the two gases produce entirely different patterns or characteristics as the result of an explosion. Natural gas, with its vapor density of 0.6, tends to rise within the structure. An explosion of it will follow a pattern that has been repeated so often as to become the typical example. Depending, of course, upon the construction of the structure, the roof will most likely be lifted several feet in the air; the walls fold out, and the roof then drops back. It may drop into the basement or onto the foundation. If the interior partitions have been left intact (still standing) by the explosion, the roof will be draped over them. Figure 6-12 illustrates this situation. The partition walls will remain standing because of equal pressure on both sides of them. Should an internal partition wall be pushed out of the vertical, the investigator can ascertain the direction from which the force moved the wall. This may indicate the direction from which the explosion occurred. Pictures are frequently left hanging in place; and, most significantly, most industrial machinery will not be seriously damaged. Most destructive damage to piping and other equipment not directly within the path of the explosive forces will be the result of such incidents as structural collapses.

Table 7-1 lists some of the properties of the various gases or vapors that are of interest to the investigator.

Ignition Velocity and Pressure Release

The ignition velocity of burning natural gas is approximately one foot per second, with from 0.1 to 0.8 seconds being required to develop maximum pressure. This will depend, of course, a great deal upon the position of the mixture within the flammability range. The intensity of any ensuing fires, as well as of

TABLE 7-1
Properties of Gases

	Flammable Limits (% gas in air)		Vapor Density (Air = 1.000)
	Lower	Upper	
Natural gas (methane)	5.0	15.0	0.6
Carbon di-(bi) sulfide	1.0	44.0	2.6
Propane	2.2	9.5	1.56
Carbon monoxide	12.5	74.0	0.97 (1.0)
Hydrogen	4.0	75.0	0.0695
Gasoline	1.4	7.6	3.0 to 4.0
Acetylene	3.0	82.0	0.899

Source: Adapted from Eugene Meyer, *Chemistry of Hazardous Materials* (Englewood Cliffs, N.J.: Prentice-Hall, 1977).

injuries to people, often reflects the point at which ignition occurs between the lower and upper flammability limits. Some authorities have observed that at the lower explosive limits, it is not uncommon to find relatively uninjured occupants crawling from the complete wreckage of a structure. In fact, some consider the danger from falling structural members to be greater than that from serious burns. As previously indicated in the discussion on fire damage, if the gas ignited at or above the point of perfect combustion, the intensity and body of heat is so much greater as to offer a very serious hazard to the lives of the occupants. Here again, the careful examination of the remaining structure, its condition, and the amount of fire damage will aid the experienced investigator in the reconstruction of the conditions under which the explosion occurred.

Because it is a pressure rather than a disruptive force, the gas explosion often creates a condition in which the person who actually triggered the explosion walks away. The janitor whose match caused an explosion in an apartment building was the only survivor of the twelve people in the structure.

Johnson indicates that, in contrast with the normally relatively slow ignition of gas, under certain (and not too clearly understood) conditions of pressure and temperature, detonation waves can be set up to raise flame speeds to such supersonic figures as 3,000 to 12,000 feet per second. In confined spaces, the localized expansion of gases may pile-pressure the as-yet-unexploded gas ahead of the burning gas to an extent that starts a detonating wave, which can then explode with tremendous force.[4]

Previous discussions have been based upon the premise that the initial pressure of the gases involved was atmospheric (one atmosphere). This would be the case in a building or structure. However, if the gases were initially under pressure higher than atmospheric, such as one would find in a transmission line, their behavior would be quite different. For example, instead of maximum pressures

[4] Johnson, "Fingerprints of Fire," p. 28.

of 115 psig, we would find maximum pressures of 2,000 psig from an initial transmission-line pressure of 500 psig. With an initial pressure of 1,000 pounds, the maximums could reach 5,000 psig.

Initial pressures of the gas above atmospheric also affect the flammability limits. Higher initial pressure will usually drop the lower limit of flammability by only about one percentage point. However, the upper limit, which is about 14 percent with a pressure of one atmosphere, will move up to about 44 percent with an initial pressure of 500 psig. An initial pressure of 1,000 psig will raise the upper limit to 53 percent, and to 60 percent at 3,000 pounds. (All figures are for natural gas.) Thus, gases become more active, and are therefore potentially more dangerous, as the initial pressure increases.

Since tanks and transmission lines naturally cannot be designed to withstand explosive forces, any mixture of gas and air in them at any time is extremely dangerous.

Rate of Energy Release

The rate of release of energy within a given period of time, rather than the amount of energy released, determines the force of an explosion. The atmosphere has the ability to readily and quickly absorb pressures and forces. To be effective, gas pressures developed by an explosion must overcome this ability of the atmosphere to absorb the forces generated. This can be done only by increasing the speed at which the pressure is generated.

Explosion forces can be released or neutralized. For many years, buildings have been designed with "weak" walls and roofs to quickly release the pressures developed by an explosion. Today, methods have been developed to neutralize or stop explosions before they can develop sufficiently to cause any damage. Explosion-suppression systems are now found in many industrial settings.

The amount of force created by an explosion is dependent upon the type of explosive. An explosive force moves in a direct path away from the center of the explosion or explosive material in a 360° pattern. Should a building be completely filled with an explosive gas mixture and then become ignited, the force may be so distributed that the entire building will be almost uniformly destroyed. A partially filled structure will tend to follow the pattern described previously as the "typical" gas explosion.

Blast waves, like sound or light waves, will bounce off reflective surfaces. This reflection may cause either a scattering or a focusing of the wave. A blast-pressure wave will lose its power and velocity quickly when the detonation takes place in the open. For example, if a block of explosive is detonated in the open, the blast wave will dissipate at a distance of 100 feet from the point of detonation. If the same charge had been placed inside a large-diameter sewer pipe or a long hallway and detonated, the blast pressure would have been still measurable at 200 feet or more. This is due to the reflection of the blast wave off the surfaces surrounding it. In fact, the reflected wave may actually reinforce the original wave by overlapping it in some places.

Since the reflected wave is a pressure wave, it will exert physical pressure. Similarly, a blast-pressure wave may be focused when it strikes a surface that acts as a parabolic reflector, just as sound waves are focused and directed into a microphone by the TV soundman along the sidelines at a football game, enabling the home viewer to listen in as the quarterback calls the signals.

Shielding occurs when the blast-pressure wave strikes an immovable object in its path. If a square, solid concrete post several feet thick is placed in the path of the blast-pressure wave, the wave will strike the post, and the post will, in effect, cut a hole in the pressure wave. The area immediately behind the post is afforded some protection from the pressure of the explosion. At some point beyond the post, however, the split blast-pressure wave will re-form and continue, but with diminished force.

When dealing with detonations that have taken place inside buildings, one can note many unusual effects that are due to reflection or shielding. These effects account for such strange things as an entire wall of the structure being blown out, while a mirror on the opposite wall remains uncracked.

When an explosive charge is buried in the earth or placed under water and detonated, the same violent expansion of gases, heat, shock, and loud noise result. Since earth is more difficult to compress than air, and water is not compressible at all, the detonation will seem less violent, but actually the energy released is exactly the same as would result from a detonation in the open air. The effect of this violence is, however, manifested in a different manner. The blast wave is transmitted through the earth or water in the form of a shock wave, which is comparable to a short, sharp, powerful earthquake. This shock wave will pass through earth or water just as it does through air, and when it strikes an object such as a building foundation, the shock wave will, if of sufficient strength, damage that structure much as an earthquake would. The entire building is shocked from top to bottom. Walls crack, doors jam, objects fall from shelves, and windows shatter. Below ground in basement areas, a strong shock wave may buckle walls inward, rupture water pipes, and heave concrete floors upward.

For example, if a 50-pound explosive charge is buried ten feet in the ground and detonated, cast iron pipes 30 feet away will probably be cracked or broken; brick, tile, and concrete sewers 40 feet away would be cracked and broken; and damage to building foundations can be anticipated for 50 feet and beyond.

An explosive charge detonated underwater will produce damage at even greater distances, because, unlike earth, water is not compressible. Since it cannot be compressed and thus absorb energy, it transmits the shock wave much faster and farther and consequently produces greater damage within a larger area.

Damage from a boiler explosion in many ways resembles that from a chemical explosion or detonation, but it is usually not as severe. The explosive energy in a boiler-type explosion is also concentrated, like that expended from the chemical-type explosion.

Natural Gas (Methane)

Natural gas has been described as being fugitive, or very difficult to contain. Since it is lighter than air, its normal tendency is to rise and to dissipate into the atmosphere. However, where its path is blocked, such as by a blacktop or concrete surface, or even by frozen ground, it habitually travels for considerable distances underground. It often finds its way into basements, sewers, pipe chases, and tunnels. Gas is known to have traveled in this manner for hundreds of feet.

Natural gas, as prepared for the market, is usually almost entirely free from odor. Since this constitutes a definite use hazard, virtually all distribution codes require sufficient artificial odorization to serve as a warning of escaping gas. The characteristic odor that most people associate with gas is artificial, usually created by the chemical mercaptan. Unfortunately, such an odorant, not being an integral part of the gas, is sometimes filtered out in transit. This can be especially true if the gas passes through the types of soil that act as natural filters.

The amount of gas involved will affect both the intensity and the duration of a gas fire. Although the actual volume of pure gas burned in a confined space is usually small, a great volume of gas may be involved in a sustained fire. The rupture of a transmission pipeline may release a tremendous volume of gas. Standard valve spacing in transmission pipelines is eight miles. Should a 24-inch line break with a gas pressure of 700 psig, 6,316,000 cubic feet of gas will be released into the atmosphere. At the lower limit of flammability (4 to 5 percent), the resultant combustible gas-air mixture would be 140 million cubic feet. Under these conditions, if the escaping gas becomes ignited, it must be allowed to burn. It is very dangerous to attempt to extinguish a gas fire unless the flow of gas can be stopped.

Similarly, 4,000 gallons of liquefied petroleum gases, when released to the atmosphere, will produce 146,000 cubic feet of pure gas. At the lower limit of flammability (2.4 percent), over 6 million cubic feet of an explosive gas-air mixture is formed.

Liquefied Petroleum Gases

Liquefied petroleum gases, such as butane and propane, are the most dangerous of the commonly used fuel gases. When compared with natural gas, they have four characteristics that make them more hazardous.

1. Liquid petroleum (LP) gases are more transient or fleeting than other gases. They are more likely to leak at ordinary pipe joints unless special gaskets and sealing compounds are used. Their ability to find a spot to leak from might be compared to antifreeze leaking from the cooling system of an automobile. When testing a system following an explosion, it is important to test with the same gas that was being used at the time of the explosion. Air and other gases might indicate that the system is tight, when LP gases will leak from it.

2. Upon release from the pressure and confines of the container, LP gases will change, almost instantaneously, into gaseous form. Most flammable liquids, such as gasoline, will remain in liquid form and slowly evaporate when released from a container. The expansion rate of liquid petroleum gases is not only rapid but at a ratio of approximately 270 to 1. However, in their gaseous form, they behave much like other flammable gases.

3. When confined within containers, their vapor pressure rises rapidly with even moderate increases in temperature. The approximate vapor pressure at different representative temperatures is shown in Table 7-2. Overfilling of LP-gas containers increases the hazard of vessel failure and possible disastrous results. The investigator must keep these facts in mind when dealing with LP-gas incidents.

4. Both butane and propane are unusual among fuel gases in that they are heavier than air. Butane has a vapor density of 2.0, propane of 1.6. This property causes these gases to seek lower levels. They will remain underground or stay close to the ground while most other fuel gases are dissipating into the atmosphere. Any leakage from an underground tank or line is likely to accumulate in low areas such as basements, crawl spaces, sewers, or manholes.

Liquefied petroleum gases, like most other compressed gases, are subject to the B.L.E.V.E (boiling liquid expanding vapor explosion) when their containers are exposed to extreme heat. This usually occurs from direct flame impingement on the container, which weakens it structurally. As indicated in number 3 above, the boiling liquid vapors of the LP gas can exert enough pressure upon the

TABLE 7-2
Vapor Pressures

Temperature	Propane	Butane
	Pounds-per-square-inch gauge	
−44°F	0	0
0°F	24	0
32°F	54	0
70°F	124	31
100°F	192	59
130°F	260	67

weakened area to cause a rupture. When such an incident occurs, the fire investigator is expected to come up with the correct answers from the investigation.

Ignition and Ignition Sources

The energy required to ignite an explosive mixture may be very small. The U.S. Bureau of Mines has indicated that when the human body is electrically charged to 10,000 volts, the weak static discharge between the finger and a metal cup of petroleum ether is sufficient to ignite the vapor rising from the cup. It also states that certain combustible gas–air mixtures can be exploded by static sparks where the energy is as low as 1/10 millijoule. Furthermore, it points out that the mere shock of a sudden pressure change has proved to be sufficient to ignite other gas–air mixtures.

Some of the more familiar ignition sources are these:

Flames: Open lights; matches and cigarette lighters; fires from boilers, furnaces, and water heaters; burning materials; incinerators

Arcs and sparks: Static electricity; electric shorts; sparks from tools, grinding wheels, and arc welding; lightning

Heated and glowing materials: Glowing metals; hot stoves; hot surfaces; cinders; electrical filaments and electric lights

Of all these, static electricity is one of the most difficult to control. There are few industrial operations in which it may not be present. It becomes more serious when the relative humidity is below 60 percent. Grounding and bonding are methods of reducing the hazards of static electricity.

Dust Explosions

Under favorable conditions, a dust explosion can occur in any industrial plant where combustible dusts are created or allowed to accumulate. Many materials that are perfectly innocuous when intact, such as most metals, become explosion-prone when finely divided to dust. Any list of potentially hazardous processes would embrace at least some part of every major industry.

Combustible solids existing in the form of very small particles or dusts usually burn with great rapidity. A mixture of dry dust with the right amount of air, when confined and ignited, may explode in much the same manner as mixtures of flammable gases and air.

In one very important way, dust-explosion hazards are more serious than those of flammable vapors. Vapors in a room or building are usually dissipated by normal air currents, so the hazard is eventually reduced or eliminated. Dust, on the other hand, tends to build up and remain on horizontal surfaces of structural members and equipment. Consequently, a very small primary explosion may dislodge this accumulation, disperse it, and place it in suspension throughout the room. If a combustible mixture exists, any source of ignition may set off a much greater secondary explosion, causing heavy loss of property and possibly of life.

Although some materials are pyrophoric (capable of igniting spontaneously when exposed to air), three conditions are normally necessary in order that most dusts can explode—the dust itself must be combustible, the dust must be dispersed in air (in proper proportions), and, with the exception of certain pyrophoric (self-igniting) dusts, there must be a source of ignition.

Almost all dusts are explosive unless their origin is nonoxidizing (noncombustible), such as sand, rock asbestos, fuller's earth, rock dust, and similar material. Another point that determines the possibility of a dust explosion is that, as is the case with flammable gases, all combustible dusts have a minimum concentration in air (oxygen) below which they will not ignite regardless of the intensity of the spark or flame.

Also, sufficient dust must be present in the air to constitute an explosive hazard. The exact amount depends upon the material involved, the uniformity of distribution in the air, the amount of moisture, and the particle size. There is probably also a corresponding upper limit of concentration above which insufficient oxygen (air) is present to support ignition. However, research on this is so complex that little tangible information is available.

PATTERNS OF DUST EXPLOSIONS

Almost all dust explosions follow a well-defined, prescribed pattern. A small disturbance accompanied by a spark or flame causes an explosion. This, in turn, torches one or more secondary events, which are fed by the dust disturbed in silos or bins, or falling from rafters, beams, window sills, or any other horizontal surface.

In many respects, a dust explosion is very similar to a gas explosion, especially in regard to the pressure generated by the *initial* blast. Most dusts have pressures similar to that of methane, or natural gas. However, *the violence of an explosion is proportionate to the speed of the release of the total energy developed*. In the case of certain explosive dusts, this rate of pressure rise has been observed at over 10,000 pounds per square inch per second. This is highly significant, since the explosive force increases as the rate of pressure rise increases.

In comparing the explosive force of dynamite with that of gasoline, we are often told that, pound for pound, that of gasoline is higher. But the explosive force is not the total energy developed. The fact is that, in dynamite, the energy is released many times faster. Thus, a wide difference can be expected between an explosion of grain dust and one of stamped-aluminum dust. The average rate of rise for grain dust is 1,000 psig per second and that of stamped-aluminum dust ten times greater, or 10,000 psig per second.

No structure could be expected to withstand the full force of any major dust explosion. However, if the rate of rise is slow enough, the pressure within the structure may be released by the quick venting of light windows, breakaway sashes, and explosion vents.

Summary

Just as with all other investigative work, the determination of the cause of an explosion is the careful assembling of all the evidence and results of the incident. These are thoroughly matched against all known characteristics of the suspected cause of the occurrence. Thus, in the case of a gas explosion, it must be shown that escaped gas was available; that it had a plausible source of entrance into the affected area; and that the final results were consistent with the effect of power pressure rather than disruptive force. In other words, all the pieces of the puzzle must fit into a single continuous sequence of events and the effects must involve no unexplainables.

QUESTIONS

1. Define the following terms:
   brisance
   deflagration
   detonation
2. Describe the three types of explosions defined in the text.
3. List the three almost simultaneous events that take place in an explosion.
4. Describe the effects of the two pressure waves developed by an explosion.
5. Why do we encounter a greater amount of fire damage when the gas–air mixture is ignited at or near the upper limits of flammability?
6. What is meant by flammability limits?
7. What is the pressure range for natural gas (methane) at the lower end of the flammable limits? at the upper end of the flammable limits? at the optimum mixture?
8. What is important about the rate of release of energy in an explosion?
9. What are the four characteristics of liquefied petroleum gas that make it more dangerous than other commonly used fuel gases?
10. What is the typical pattern of a natural-gas explosion in a structure?
11. What is the significance of vapor density in connection with combustible gases and vapors?
12. What is the typical pattern of a dust explosion?
13. The violence of an explosion is proportionate to what?

The Fatal Fire

Aside from the cause of the fire, whenever a fire death occurs, the fire investigator must seek the answer to four additional questions. The questions deal with:

1. The identity of the victim or victims
2. Determination of the cause of death
3. Determination of the manner of death
4. Determination of the approximate time of death

Anytime there is a sudden, violent, unexpected death, there is the need for an autopsy. The autopsy should be performed by a pathologist skilled in the investigation of deaths involving violence. If possible, the services of a fully trained forensic pathologist should be used.

Forensic Pathology

There are two major branches of pathology, anatomical and clinical. The anatomical pathologist has been trained to deal with the examination of tissues and organs at autopsies or from surgical procedures. He or she will make diagnoses to identify and recognize disease and injury patterns with the naked eye or the microscope. The clinical pathologist handles the rest of laboratory medicine—performs examinations, supervises, and teaches people to do other tests done in a hospital laboratory. These people are trained to deal with natural disease.

One subspecialty within the field of pathology is known as forensic pathology. "The forensic pathologist uses the techniques and tools of medicine in general, and the specialized tools of pathology in particular. He applies those tools and techniques to the investigation of those matters involving medicine

110

that are likely to come before a court of law. It is the pathology of the courtroom."[1]

Fire deaths are very complicated from a medical point of view. Few pathologists have been trained in the investigation of sudden, violent, unexpected deaths, such as those resulting from fires. There is a great need for close cooperation among the fire investigator, the coroner or medical examiner, and the forensic pathologist whenever bodies are recovered from a fire. It is also extremely important that the investigator be able to help the pathologist obtain the information he or she needs before performing the autopsy.

Just what is this information that should be available to the pathologist? Information should be made available about the supposed victims, anything that might assist in identification: medical records, dental charts, records of industrial accidents, and X rays, if at all possible.

Identification of the Victim or Victims

The first and most important question that must be answered is that of the identity of the dead body or bodies. This can sometimes be very difficult. Several methods may be used for this purpose. The least reliable method of identifying a dead body, whether burned or not, is by direct visual examination, sometimes called "gross examination." People just don't seem to be able to identify properly under the extreme stress of death. Sometimes relatives or friends refuse identification because of the condition of the body. Two other means of identification, which are considered fairly unreliable, are by clothing and by personal effects—such things as jewelry, lockets, rings, religous medals, rosaries, keys, pocket knives, and so on.

There are at least seven methods of identification based upon a comparison of the deceased body with known characteristics of possible victims. These are all considered to be more reliable than the three mentioned above.

1. *The sex of the body.* This may be easily determined with any body that has not been incinerated. In large disasters, with many bodies, the first separation is made by sex. At the Texas City disaster, where hundreds of unidentified dead created a tremendous problem, after separation by sex, the male bodies were separated into circumcised and uncircumcised groups. This was found to be of considerable help in reducing the number of bodies that had to be viewed for identification. In the case of badly burned, almost destroyed bodies, sex may still be determined by an examination of the torso by the pathologist. The deep-seated organs in the pelvic region, the prostate of the male and uterus of the female, are usually among the last organs to be destroyed by the fire.

[1] Laurence R. Simson, Jr., M.D., "Fatal Fire Investigations," talk delivered at Thirteenth Annual Wisconsin Arson Seminar, La Crosse, Wis., June 7, 1978.

2. *Fingerprints.* Probably the most accurate and positive means of identification is by fingerprints. The national fingerprints file, established by the FBI in 1924, contains millions of fingerprints that can assist in identification. The laboratory technician can assist in taking the fingerprints. If there is a glove-like separation of the epidermis from the hands, this may be used. If necessary, the hands can be removed and submitted to the laboratory for study and obtaining of fingerprints.

3. *Dental features.* Teeth have individual characteristics that provide another reliable method of identification. The teeth of the victim may be compared with dental records and dental X rays. From the teeth other information regarding age, race, preexisting disease, habits, and in some cases occupation can be obtained. Examination should be made of dentures for name or identification number, which may be affixed to the denture base. Dentures made while the person was in service may contain the owner's GI serial number or Social Security number. It may be necessary for the pathologist to remove either the lower or upper jaw, or both, during the autopsy to facilitate examination and identification of the teeth.

4. *Skeletal features.* Bones resist the effects of environmental conditions, time, and heat. Assistance may be obtained from anthropologists. They are trained to determine such things as the age at death, sex, race, and evidence of prior disease or injury; to estimate the height by an examination of the skull and other bones; and to distinguish human bones from animal bones.

5. *Serologic and cytologic studies.* These may be used to determine blood group and Rh type, as well as other factors, such as whether the sample is animal blood or human blood. The identification of species, Gm factor, sex chromatin, and karyotyping is also possible.

6. *Autopsy examination.* A postmortem examination may reveal such things as occupational scars and marks, tattoos, evidence of preexisting diseases or prior injuries, congenital defects, operative scars, and the absence of organs owing to surgical procedures. These may be compared with the medical and employment records of possible victims.

Occupational scars would include missing fingers or parts of limbs incurred in industrial accidents. When the burned body is missing fingers, hands, or portions of limbs, a question arises as to whether they were previously amputated or burned off during the fire. An autopsy can usually answer this question, as well as identifying other prior surgical procedures such as an appendix operation, hysterectomy, mastectomy, or orchiectomy.

As tattoos fade over the years, the ink is absorbed into the body. It may be deposited in the lymph glands in the armpits. For example, a victim's right arm has been burned off just above the elbow. One of the possible victims had been tattooed on the right forearm. The pathologist may be able to find traces of the ink deposits in the lymph glands of the right armpit. This would indicate the corpse had been tattooed.

Tattoos can be extremely important; the last body to be identified in the Texas City disaster was finally identified through a tattoo.

7. *Radiography*. Films taken during life can be compared with postmortem films. One that may be important in this category is the chest X ray. In addition, postmortem X rays will reveal foreign materials and metallic fragments not observed during the autopsy examination.

Other items that may assist in the identification process and should not be overlooked are eyeglasses and contact lenses. The eyeglasses, together with the frames, may be compared with medical records. Contact lenses should be examined with ultraviolet light for markings of the manufacturer.

Determination of the Cause of Death

The principal cause of fire deaths is carbon monoxide poisoning. Rarely do fire victims die from burns; most thermal damage to the body is postmortem. Where there is fire, there is smoke; where there is smoke, there are particulate material, carbon monoxide, and soot.

Carbon monoxide (CO) is very poisonous. Two pounds of cotton waste, when burned, is sufficient to develop a rapidly lethal level of CO in a good-sized living room. CO has two different mechanisms to affect the body. First, acute or high concentrations affect the ability of the blood to carry oxygen (O_2). The carbon monoxide combines with the hemoglobin of the blood to form carboxyhemoglobin (HbCO). Hemoglobin has between 200 and 300 times a greater affinity for CO than it does for oxygen. Blood cells that are loaded with CO cannot transport the oxygen the body needs. As a result, the body becomes oxygen-starved. The second mechanism involves long exposure to lower levels of CO. In this circumstance, not only are the red blood cells loaded with CO instead of oxygen, but the CO directly poisons tissue cells.

During the postmortem examination, blood for analysis of the amount of CO may be taken from the heart. If the laboratory analysis reports carbon monoxide content, the following percentages apply:

Under 10%: Normal background.

15–18%: The person was alive in a CO-contaminated environment.

Levels in excess of 45% are lethal and indicate that the person died as the result of carbon monoxide asphyxiation.

60–70%: People who die during a house fire will be likely to absorb this amount.

In the event that there is a dead animal found in the house with the body, have blood drawn from the animal to check the CO level. If someone comes into the house to kill the occupant, and the occupant has a dog, most of the time the dog will have to be killed first. If there is no CO level above the normal background in the dog's blood, the dog was dead before the fire.

The forensic pathologist will often find soot in the stomach of the victim. If a person was alive during the fire, soot will be found in the airway—that is, in the nose, throat, larynx, trachea, and bronchi. To make the determination, a neck dissection is required. It is important that the pathologist examine the airway as high up as he can get it out—to the base of the skull.

The absence of any carbon monoxide in the blood of a dead body found in a burned structure would be strong circumstantial evidence that the person was dead when the fire began. The principal question raised the determination of the cause of death is *Was the victim alive during the fire*? There are occasions when the fire victim will die from the fire without any carbon monoxide being present in the blood or soot in the airway. This occurs when the victim is hit in the face with an extremely hot fire, such as the blast from a flammable-liquids explosion, the backfire from a furnace, or the fire ball from a rear-end automobile accident.

POSTMORTEM THERMAL DAMAGE

There are a number of signs of postmortem thermal damage. Some of these are commonly found in connection with fire deaths. For instance, the body contains fatty tissues from which the fat will be rendered. It is the burning of this fat and grease that destroys the corpse. Whenever this occurs, the intense heat may produce long splits in the skin that can resemble lacerations. Another common sight involving burned bodies is a split-open abdomen, with the intestines sticking out but unburned. It is the result of the intestines filling with gas, which then expands from heat. As the body cools, the intestines break out of the weakened stomach walls. This is a postmortem artifact, or changed appearance or condition resulting from heating of the dead body, not the result of someone's using a knife or razor on the victim.

When the corpse is incinerated, the first parts of the body usually destroyed are the extremities, such as the fingers, hands, or feet. It is not unusual for the scalp to be destroyed if the body is badly burned. The skull can be fractured by the heat, as the result of steam generated inside the skull. Fractures of this type usually follow the suture lines of the skull. The bone is pushed out, not depressed as it would be if the fracture was caused by a blow. In this case there might also be lines radiating out from the point of impact. Flames in contact with the skull can cause extensive charring. This can be in the form of a brownish-red deposit between the skull and the covering over the brain. It looks like a heavy mass of blood, and it is caused by the tissue juice being cooked out of the bone (skull). This too is a postmortem thermal artifact, but it can be mistaken for antemortem injuries and erroneously interpreted as evidence of a crime.

When broken bones are encountered, the pathologist must determine if the fracture was produced by heat alone, or by some outside agency. Bones crack in a fire; but postmortem fractures can occur in connection with structural collapse. Fractures found at the base of the skull are not produced as a result of the "steaming" of the brain. Their cause must be determined.

When the body has not been incinerated but has been burned, a determination must be made as to whether the burns are antemortem or postmortem. This can be very important. Arson is a fairly common method of covering a homicide, but burns do frequently cause death, particularly when flammable liquids are involved. It is usually the secondary effect of the burns that actually causes the death. A burn is a violent insult to the body, affecting all the body systems. The burn will free toxic substances that may be absorbed into the blood. The shock that the body experiences after a burn is the cause of many fire deaths.

The pugilistic position that many bodies assume in a fire means nothing. To the uninformed, the pugilistic position demonstrates the victim's attempt to defend himself. This is not true. This position is the result of the contraction of the larger muscles from the heat of the fire. The posture in which a body is found in a fire death has no connection with its antemortem posture.

It is so easy for local authorities to say, "This fire looks like an accident," and then to forget the whole thing. The body is embalmed and buried without an autopsy. But what happens when, three months later, a member of the family comes in from out of state and says, "We think so-and-so killed Daddy"? Where is the investigator at this point? With no autopsy, how can he prove there was no foul play?

Determination of the Manner of Death

The next major problem raised by a fire death is the manner of death. There are four categories of death involved in this problem: Was the death from *natural* causes, was it an *accident*, was it a *homicide*, or was it *suicide*?

NATURAL CAUSES

The category of natural causes includes such deaths as those resulting from a heart attack, chronic disease, or any other natural disease process. After the death, the fire occurred, and then the body was discovered. These circumstances might lead to the conclusion that the death was the result of the fire. An autopsy and a thorough postmortem investigation will establish whether the death occurred before the fire and if it was from natural or accidental causes.

ACCIDENTAL CAUSES

The vast majority of fire deaths fall into the accidental category. The circumstances of the fire and the autopsy report will aid in establishing the true cause of death and the true cause of the fire.

Careless smoking, careless disposal of smoking materials, and misuse and careless use of flammable liquids—especially gasoline—are the most common types of acts leading to accidental fires and fire deaths. Faulty electrical, heating, and cooking equipment, as well as poorly maintained chimneys and fireplaces, all cause accidental fires.

Almost any fire can cause death or injuries. The various accidental fire causes have been covered in detail in Chapter 6. As indicated previously, the vast majority of accidental fire deaths are caused by carbon monoxide poisoning.

SUICIDE

Few people commit suicide by burning themselves. But there have been a number of protestors using this method in recent years. They douse their clothing with gasoline and then intentionally ignite themselves. On the other hand, an attempted suicide by gas can, as we have seen, result in an explosion and fire that does kill the individual.

HOMICIDE

When the origin of the fire is incendiary, any fire death resulting from it becomes a homicide. A 14-year-old boy wanted to ride his neighbor's motorcycle, but the neighbor refused to permit it. To "get even" with the neighbor, the juvenile poured gasoline over the motorcycle in the neighbor's garage. He then threw a match on the motorcycle. The resultant fire emerged from the garage door and went up into the bedroom window directly above it. Two children, aged 3 and 4, died in their beds. Homicide was not the intent; homicide was the result. Figure 5-17 shows the burn pattern out of the garage-door opening (covered with plywood) and into the window of the bedroom.

A fire is often used as a means to hide a homicide. Again, the absence of carboxyhemoglobin in the blood and the absence of soot in the airway indicate a death prior to the fire.

Determination of the Approximate Time of Death

A determination as to whether a fire death occurred prior to or during the fire is important in establishing the cause and the manner of death. The three criteria that the forensic pathologist uses in making this determination are those previously discussed in determining the cause of death. They are the carboxyhemoglobin content of the blood, the presence of soot and combustion particulate in the body airway, and the appearance of skin burns.

Dr. John F. Edland summarized the procedures of the forensic pathologist and the fire investigator in connection with fire deaths as follows:

In an investigation of deaths of persons recovered from fire, the following points should be emphasized:

1. Following a fire, a thorough investigation should be conducted to ascertain the possibility of the presence of unsuspected victims.

2. The circumstances surrounding the fire should be carefully investigated by the medical examiner and law enforcement agencies, such as the police, fire marshals, etc., and all this information should be given to the forensic pathologist prior to the beginning of the autopsy.

3. When the presence of human remains is known or suspected, the medical examiner should be notified immediately so that he or his deputy may proceed to the scene.

4. Adequate photographs and sketches of the scene should be made.

5. A thorough, extensive, complete, and unhurried postmortem examination should be made by an experienced pathologist having the necessary equipment and physical facilities for such an examination.

6. Finally, the most valuable tool in the investigation of the dead body recovered from a fire is a high index of suspicion and unlimited curiosity in the minds of all those engaged in the investigation.[2]

QUESTIONS

1. What is a forensic pathologist?

2. Why are the services of a forensic pathologist so necessary in the case of fire deaths?

3. What are the four questions that must be answered when a fire death occurs?

4. What is meant by "gross examination"? Why is it considered the least reliable method of identification of a dead body?

5. Name six methods of identification of dead bodies considered to be fairly reliable.

6. What is the most accurate and positive means of identifying a dead body?

7. What is carboxyhemoglobin (HbCO)?

8. What percentage of carbon monoxide in the blood is considered to be normal background level?

9. What should the forensic pathologist look for in the airway of the fire victim?

10. The "pugilistic position" so often seen in connection with deaths:
   a. indicates the victim died defending himself from attack.
   b. strongly indicates the possibility of a homicide.

[2] John F. Edland, M.D., "Fire Victims," in *Forensic Pathology—A Handbook for Pathologists*, eds. Russell S. Fisher, M.D., and Charles S. Petty, M.D. (Washington, D.C.: U.S. Government Printing Office, 1977), pp. 109–110. (Preparation of this document was supported by a grant awarded by the National Institute of Law Enforcement and Criminal Justice to the College of American Pathologists.)

    c. means nothing at all.

    d. will nearly always be followed by sudden convulsive jerks.

11. What are the four categories involved in the manner of death?

12. What does Dr. Edland say is the "most valuable tool in the investigation of the dead body recovered from a fire"?

9 Care and Handling of Physical Evidence

Evidence is not an item or thing. It is not a category into which the investigator can insert some items and omit others. Evidence is not measured by what it is, but rather by what it does. It must do something within the framework of the case, or it is not evidence. Evidence may establish the *corpus delicti* of the crime of arson, or it may establish an accidental cause of the fire.

As indicated in Chapter 3, most arson cases are proved by circumstantial rather than direct evidence. Much of this circumstantial evidence is the physical evidence used to establish the *corpus delicti* of the crime of arson.

Evidence may be the account of an eyewitness who saw someone or something. It might have been dark black smoke at the start of the fire. The odor of fuel oil or gasoline recognized by the firefighter can be evidence. On the other hand, evidence might be items that show the fire was not incendiary.

Evidence can establish either guilt or innocence. Through evidence, the identity of the criminal can be established. The eyewitness sees someone fleeing the scene prior to the fire. Or someone watches the suspect carry gas cans into the structure. These cans can be traced to the offender, or a gas-station attendant remembers filling them for the suspect. Evidence can be a statement that places the suspect miles away from the scene when the fire started. The records of the business involved in the fire, or any fingerprints found on the scene—each can be evidence. In short, evidence can be anything that establishes facts.

When evidence is measured by what it does, it must fulfill certain requirements:

1. Does the evidence establish the elements of the crime?
2. Does the evidence establish the identity of the offender?
3. Does it tie the offender to the scene of the crime?
4. Does the evidence establish the necessary mental state of the criminal, or the mental state necessary to commit the crime?

5. Does the evidence support or destroy statements made either by witnesses or by the suspect?

If the material being considered will do one or more of these, it is evidence.

Physical Evidence

Physical evidence is evidence that has a physical substance or existence—something that can be perceived by the five senses. In an arson case, physical evidence may be a burn pattern that indicates a criminal fire, along with the photographs of the pattern. It may be properly collected samples from the fire scene and the laboratory report indicating the presence of an accelerant, such as a petroleum hydrocarbon.

Physical evidence is used to establish or assist in establishing the incendiary nature of a fire. It may be a burn pattern, ignition device, trailer, plant, or container of accelerant. It may be anything that is associated in any manner with the ignition, origin, or spread of the fire. Ignition devices can include such things as candles, alarm clocks or timers, wicks, matches, fusees or flares, blasting caps, short-circuited electrical devices, and explosive and chemical igniters and devices. The trailers may be flammable or combustible liquids, together with paper, cloth, curtains, carpeting, and other combustible materials. These burn from the igniter or ignition point to the plant or main body of the initial fire. Trailers are the devices that assist in the spread of the fire. They usually leave distinctive burn patterns, as seen in Figures 5-19, 6-16, and 6-17.

> The fire investigator should remember that without witnesses, the physical evidence becomes the only witness at a trial. Such physical evidence will speak convincingly for itself, but it must be handled competently, from its discovery at the crime scene until its presentation at the trial.[1]

That the evidence was obtained legally is not at issue in this chapter, but rather how it is handled, packaged, and labeled, and in what condition it is delivered and introduced into court. The first consideration is the "chain of custody" or "chain of possession" of the evidence. This chain must take each item of evidence from the place it was discovered or taken possession of to its final presentation in court. In numerous cases, the courts have held that there must be an unbroken chain of possession established between the article seized and the article analyzed by the laboratory. This chain must continue from the laboratory until it is introduced in court.

All fireground evidence must be photographed in place, where it was found

---

[1] Kent A. Oakes, "The Role of the Crime Laboratory in the Prosecution of an Arson Case," paper delivered at Twelfth Annual Arson Seminar held by the Wisconsin Department of Justice, Madison, Wis., June 8, 1977, p. 3.

prior to being moved or disturbed. Such photographs are taken so that the investigator will be able to remember where the evidence was located and be able to prove that it was at the fire scene. After the photographs have been taken, the evidence is collected and placed in its container. The area is then rephotographed to show that the evidence in the first pictures has been removed from the area. In addition this second series assists in establishing the time of the search by the investigator. This method tends to prevent suggestions that the item was picked up at some later time, upon the investigator's return to the crime scene. The importance of this timing was discussed in Chapter 3, in connection with *Michigan* v. *Tyler* and *People* v. *Tyler and Tompkins*.

Containers for Preserving Evidence

To use a container that just gets the job done marginally is worse than using no container at all. The marginal container leaves the purity of the evidence open to attack and may cause it to be discredited. For samples to be subjected to analysis for accelerant residues, one-quart or one-gallon paint cans are the best containers. These cans should be purchased from the manufacturer uncoated. The new cans are then tested by the laboratory and certified to be free from contaminants. Such cans are preferred for the following reasons:

1. They do not leak or admit outside sources of contamination.
2. If dropped, they will not shatter as will mason jars.
3. They can be safely heated in an oven during testing by the laboratory.
4. Once sealed, they do not need to be opened for testing. The chemist can make a hole in the top to draw off vapor samples for testing.
5. The can will accept permanent markings etched into it as to what is inside, who found it, where it was found, and what case the material relates to.

Unlike used containers, the new cans are not open to attack as having been contaminated. They provide uniformity in the evidence collection, as well as being far more impressive than a collection of coffee cans and cottage-cheese containers. This presents a professional image to the court.

Documents should be enclosed in plastic-sheet protectors sold in stationery stores. Prepared this way, they are easily handled and less likely to be damaged. When properly sealed as evidence, they are not as likely to be subjected to attack for having been altered.

Small trace items, not to be subjected to analysis by the laboratory, can be placed in slide-type evidence boxes. These boxes come in assorted sizes and have no writing on them. They are usually white, are easily marked and sealed, and will do a good job of holding and preserving evidence.

Small glass vials with screw-on caps can be used to hold small pieces of evidence. This type of container offers the opportunity to view the evidence without unsealing the container. These vials can also be used in the collection of samples of liquid accelerants. The method used can be very simple.The evidence kit should contain several (about ten) glass vials with a Q-Tip in each, trimmed to fit. One bottle is used for each sample and one as a control. (It is not necessary to submit more than one control bottle per case.) The control bottle is used to show the purity of the bottles and the swabs. This is to verify that there was no accelerant present in the bottle prior to the collection of the suspect liquid. The control bottle is sealed and marked by the investigator. He then uses the swabs, one to a bottle, to collect the samples. As the samples are collected, the bottles and swabs are sealed and marked with the proper date, location, time, investigator's name, and what is in the bottle—for instance, suspected liquid accelerant. Remember, it is unnecessary to furnish a gallon of a suspected accelerant for the laboratory to work on. The more liquid submitted, the greater the fire hazard being created.

One of the most versatile parts of the evidence kit is a good selection of plastic bags. The zip-lock type is the most useful. These plastic bags should range in size from the smallest, about the size of a cigarette pack, to the largest available. Plastic bags should not be used to collect liquid samples or samples containing volatile fluids. The vapors will migrate through the bag, and the sample will be lost or contaminated. However, these bags are excellent for collecting parts of timing devices, wire, soil samples, cloth, lint, tools, or paper items.

The strength, durability, and transparence of plastic bags give them an advantage over the paper or cardboard type of containers. In addition, their uniformity and overall appearance add greatly to the professional stature of the investigator. This can be very important in the courtroom.

Collecting the Evidence

The investigator must wear a pair of sturdy rubber gloves when collecting any evidence. First, the gloves protect the hands from the effects of the accelerants or any caustic chemicals found in the fire debris. Second, they prevent exposure of the hands to infections, cuts, or the entrance of foreign materials into any open wounds or cracks. Third, they preserve the evidence from contamination with the investigator's fingerprints.

These gloves should be of either the disposable type or the heavy-duty type used by firefighters. The kind of rubber gloves generally used by the housewife are too expensive to dispose of and not durable enough to be reused a number of times. If the gloves are not of the disposable type, they should be washed thoroughly after each fire. This prevents cross-contamination of cases and, in addition, makes the gloves last longer.

To collect samples of wood, wire, pipe, and other materials for evidence, the investigator will need a kit of hand tools and a tool box to transport them. In addition to those items listed in Chapter 5, the following are found in the authors' investigator's kits and have proved useful on more than one occasion. This is a basic tool list; each investigator will want to add to it.

1   tool box (drawer type works best)
6   screw drivers (assorted flat blade and Phillips)
2   pairs channel-type pliers
1   pair slip-joint (regular) pliers
1   pair electrician's (lineman's) pliers
1   pair diagonal pliers
1   pair needle-nose (long-nose) pliers
2   hemostats (medical type)
2   small pipe wrenches
1   carpet knife with set of replacement blades
1   hacksaw with replacement metal-cutting blades
1   long-bladed knife (sheath knife preferred)
1   reversible brace
2   adjustable brace bits
1   small pry bar with nail puller
1   8-oz. claw hammer
1   small penlight
1   regular flashlight
1   roll of plastic electrician's tape
1   spool 25-lb.-test monofilament fishing line
1   putty knife (to pick up samples)
1   tablespoon (to pick up samples)
1   carbide-tipped metal scriber, preferably with a magnet on the end opposite the scribe (for marking evidence)

Most of the items listed are self-explanatory in nature and use. However, the way some are used may need further explanation.

The hemostats are medical instruments resembling small pliers that lock together at the handles. They cannot be opened without applying pressure at the handle to release them. These are very handy for picking up small items or items that warrant extra care to prevent contamination.

The brace and bits are used to drill holes. It is often necessary to collect samples from wooden floors, tables, shelves, and other porous materials. The investigator can either break these items up with an axe or use a saw. When using this method he or she must try to stuff the resulting chunks into a paint can. By use of the brace and bit, the material to be sampled can be bored into, creating shavings and sawdust. The putty knife is then used to lift the sawdust and shavings from the hole and place them in the can. This method of extracting samples offers several advantages. First, the wood is sampled from surface to base. Second, samples from more areas can be taken and will fit into cans. And third, in this form, the wood gives up any accelerant more easily in the laboratory, so that the chemist's job is less difficult. The results then delivered to the investigator permit him to do a more efficient job.

In addition to the collection of shavings and sawdust from floors, tables, and shelving, some other items should be collected when it is possible that there is contamination by an accelerant. These include samples of tile, clothing, bedding, carpeting, plastics, curtains, upholstery, plaster, rags, empty or partially empty containers, broken bottles, soil, and any metallic residue or similar residue that might be from a chemical agent.

The types of accelerants that the laboratory may be able to identify are such fuels as gasoline, fuel oil, paint thinner, alcohol, lacquer thinner, brake fluid, turpentine, charcoal-lighter fuel, linseed oil, lighter fuel—in fact, almost any flammable liquid that might be used.

The roll of tape and the fishing line are useful to secure pipes and wire out of the way when working at the fire building. They may also be used to hold conduits and wires in place while the investigator is trying to trace an electric circuit.

Sealing and Marking the Evidence

Improperly sealed or marked evidence is useless. It is not much better than no evidence at all. If the items are not sealed and marked in such a way as to eliminate attack from the defense attorney, they may not be admitted into evidence at the time of the trial. And without physical evidence, the chance of a conviction is very slim.

The following items are useful in sealing evidence:

1. Evidence tape, printed with the name of your organization and the word *EVIDENCE* in a contrasting color, plus a place for the date, time, and complaint or case number. This type of tape provides a positive seal that is tamperproof. In addition, its use adds to the professional image of the investigator.
2. Twelve-inch lengths of 28-gauge wire. These can be used to tie plastic bags closed or to attach tags to the bags. Additionally, when inserted through the plastic bags, they provide a tamperproof seal that cannot be removed without leaving evidence of the removal. Seals similar to those used by fire-extinguisher servicemen can also be used in this manner.

3. Nylon filament tape (Scotch Brand Filament Tape, pressure-sensitive Scotchpar film, #880, manufactured by the 3M Company), one inch wide. This is a very strong tape that cannot be torn and has excellent adhesive properties. It is useful on boxes or envelopes used for shipping. The evidence tape should be placed on top of this tape.

4. The type of sealer used to seal food in plastic bags. This is one of the best methods available for sealing evidence in plastic bags. It is available to the housewife and requires no modification to adapt it for evidence work. It is excellent for papers or for chemicals that need to be tested. It does not lend itself to the collection of accelerant samples. Note, when using the sealed bag, that the evidence identification tag should be placed inside the bag before sealing.

Some methods and tools useful in marking the evidence are these:

1. Broad-tip felt marking pens that will write on almost any surface. These can be used to mark paint cans and glass bottles.

2. Regular ballpoint pens. These should be included in the kit, since your own pen can be forgotten or lost, or, as is the usual case, will decide to stop working just when you need it.

3. A carbide-tipped metal scriber. This is the type used in metalworking to scribe lines on steel. It should be used to mark the paint cans prior to writing on them with the felt marker. It may be important to use both the scriber and the felt marker. The defense may challenge the felt-tip writing because it can be wiped off with solvents or wear off from normal handling. By the use of both tools, the court and jury can be shown the scribings under the writing. This will accomplish two things: first, the investigator's professional image will be enhanced, and second, the defense can no longer attack the evidence on this point. The investigator has scored a mental victory over the defense. An attempt to discredit him and his evidence has failed. The scriber can also be used to put identifying marks on other metal evidence. For a slightly higher price, the investigator can purchase a diamond-tipped scriber, which permits markings on glass in addition to metal.

4. Evidence tags. Like evidence tape, these should be printed and should contain the same information. Gummed labels may also be used. Printed, self-adhesive tags and labels are the best choice. After licking ten or twenty gummed labels, the investigator longs for the self-adhesive type and a drink of water.

The evidence tags to be attached to the wire should be about the thickness of shirt cardboard and, as previously mentioned, printed to include all the items found on the self-adhesive type. In addition, a place for item numbers might also be included—for example, 1 of 6, 2 of 6—and they should be punched for wire attachment.

To further increase the usefulness of these tags, they should be color-coded. The color will indicate how the evidence is to be handled. The following is suggested as a color code for these tags and their use:

Red: to be sent to the laboratory for processing. Write on the back of the tag the type of processing needed, such as, "Test for an accelerant," "Process for fingerprints," "Check handwriting," "Tool mark examination," etc.

Green: to be held for court in the property-control section, or to be held for expert examination. Used on company books that are to be checked by an accountant, or electrical items to be examined by an electrical engineer.

Yellow: to be held for safekeeping and may be returned to owner upon arrangements by the investigator. Used when items of value, such as money, watches, or rings, have been recovered at a fire scene and are not needed as evidence.

Just what does the investigator record on the tag? Whenever there is room, record as much identifying data as is available. With printed tags, space for the desired information will be provided. If there is limited space, or the tag is blank, some type of identifying number, such as a badge number, case number, or computer number, should be used.

When using tape, the investigator's initials should be placed at the points where the tape begins and ends, as well as at any points where one strip of tape crosses another. Mark all envelopes and plastic bags at their sealing points. Mark all cans in at least two separate points on the top where the lid meets the can. The markings at each point should show on both the lid and the can. This will assist in maintaining the integrity of the evidence. The two separate points should not be directly opposite each other.

Inventory of the Evidence

As the evidence is collected and marked, it should be entered on some type of inventory sheet. This inventory sheet should be at least in duplicate and should carry the same or similar data to that found on the evidence tag. With duplicate copies, one copy can be left at the scene or with the person in control of the scene as a receipt of the items of evidence taken. In some states, this is a legal requirement. In all cases, it will improve the admissibility of the evidence by establishing its existence at an early stage in the investigation. The inventory also establishes a permanent record of the evidence. This provides the investigator and any others working on the case in the future with a list of the available evidence at hand.

Transfer of Evidence

Once the evidence has been collected, packaged, marked, and inventoried, the next step is to get it to the laboratory. The chain of custody must always be maintained. This is simply an accurate record of who had the evidence, the date and time possession took place, and the date and time control was passed on to the next person. The fewer people in the chain of custody, the simpler the court presentation will be. The investigator can expect a good defense attorney to have every person involved in the chain of custody testify. He will hope to establish some defect in the chain. If the chain of custody can be broken, the evidence will become inadmissible in court.

If at all possible, the investigator collecting the evidence should personally take the evidence to the crime laboratory and obtain a receipt for it. In some cases, it is not possible to deliver the evidence in person; the items will have to be shipped or mailed. This is a very exact process and should be established by all receiving laboratories and submitting organizations. Where an organization finds itself in this position, it should, at the first opportunity, establish the procedure to be followed. Do not wait until the evidence is in hand and has to be shipped. If arrangements are not made ahead of time, it is almost certain that something will go wrong to delay or ruin the evidence.

Laboratory Results

The evidence has been legally and properly collected, packaged, marked, and transported to the laboratory with the chain of custody intact. What can the investigator expect to get back from the laboratory?

The well-equipped and properly staffed laboratory can work what will seem like miracles. It can take a sample of burned material and determine if an accelerant is present. In many cases, the specific kind of accelerant can be determined. It can make comparative analysis of such physical evidence as soil, blood, paint, hair, glass, fingerprints, and handwriting. Among the new methods of developing fingerprints is the laser unit, currently being used by several police departments.[2] With the laser unit, fingerprints can be lifted from wood, paper, cloth, and in some cases, human flesh, in addition to all the other surfaces that normally retain fingerprints. In some cases, the laboratory can analyze the ink used to write a suspected item and can tell the investigator the type of pen used or the company that manufactured the ink. The laboratory can match marks made by a tool against the suspected tool. It can take partially burned matches and compare the torn ends with the stubs in the matchbook. The laboratory can

[2] The laser unit is currently being used by the Montreal, Canada, Police Department and the DuPage County, Illinois, Sheriff's Police.

also match the ends of tape and rope against a suspect roll and establish whether the suspected item came from the specimen submitted. And in all these cases, if any of these articles match, the laboratory can furnish the necessary expert testimony in court.

If the investigator does his job properly in the collection and submission of evidence, the laboratory can help establish the guilt or innocence of the suspect. However, the laboratory is not a substitute for a good investigation. It is merely one of the tools used by a good, professional investigator. The better the job the investigator does, the better those tools will function. Ultimately, they will assist in bringing the investigation to a successful conclusion.

QUESTIONS

1. Describe how you would mark tape used to seal evidence.
2. You find evidence of a suspected accelerant pattern on the hardwood floor in a hallway. What is the best way of obtaining samples of the flooring for laboratory analysis?
3. You have photographed an incendiary plant. You collect it for analysis by a laboratory. Why is it important that you rephotograph the area after removing the evidence?
4. Describe some of the advantages of using new paint cans for the collection of physical evidence to be subjected to laboratory analysis.
5. Why can physical evidence be so important to an arson investigation?
6. Define the term *chain of custody* of physical evidence.
7. Discuss the importance to the defense in an arson case if it can be proved that the chain of custody of the evidence was broken.

10 Motives for Firesetting

Arson is a very personal crime. It is also a very secretive crime. It can be committed without confronting the victim or victims. The crime is personal in that frequently the victim can provide the investigator with both the motive and the suspect. Although the victim may not be responsible for the crime, he may have information of vital importance to the investigator. It is not unusual for the perpetrator of the crime to be in some manner closely associated with the victim.

Motives for arson are very similar to those of other crimes. But because arson is a crime of secrecy, the motive may not be as obvious as in other crimes. Probably the most common motive for arson today is the profit motive. Most of the time, this profit motive is designed to defraud the insurance company. Over-insurance or a sudden increase in insurance coverage may indicate that a fire is being considered. Other motives for arson include incendiarism in industry, revenge and spite, the desire to conceal another crime, jealousy, intimidation, juvenile deliquency, insanity, sabotage, riots, radical terrorism, and suicide.

The principal study on pathological firesetting available to the fire service is the one made by Lewis and Yarnell and published in 1951.[1] These researchers studied the cases of nearly 2,000 firesetters. They pointed out that at least half of those studied "have been in trouble with authorities for one or many types of anti-social activity, ranging from petty stealing to manslaughter."[2] They found "no instance of a man resorting to firesetting while doing work in which he was interested and felt satisfied."[3] Indeed, both pyromaniac and revengeful firesetters had recent dismissal as one of the precipitating causes of the crime.

---

[1] Nolan D. C. Lewis and Helen Yarnell, *Pathological Firesetting (Pyromania)*, Nervous and Mental Disease Monographs (New York: Coolidge Foundation, 1951).

[2] *Ibid*., p. 41.

[3] *Ibid*., p. 39.

Fraud and Profit Motives

One of the most common methods in arson for fraud or profit is for a group of people to enter into a conspiracy for this purpose. For instance, one of the group purchases some property for a given amount and insures it for as much as possible. Then the property is sold to another member of the group or to a fictitious person at, say, twice the original price. The insurance is increased based upon the new purchase price, even though no work has been done to improve or maintain it. The property now has a higher value, and after a period of time, it is resold at a still higher figure. The insurance is again increased to reflect the new price. After the property has been sold and resold a number of times, with a substantial increase in the insurance each time, the group decides to "sell" it to the insurance company. A fire occurs, and the building is a total loss. The property may still be worth only the original price, but the insurance value is many times that figure.

There are a number of possible reasons behind the business fire. Perhaps the owner of the business is unable to meet financial obligations such as taxes, mortgage payments, accounts past due and payable, or notes past due and payable. Or the owner wants to liquidate an unprofitable business or dissolve a partnership; no longer wants the property but can't sell it. Or he or she has too high an inventory of obsolete merchandise, no ready market for stock, or a seasonal inventory that has not moved. Sometimes orders for merchandise are canceled by the jobber, and the manufacturer is unable to dispose of the stock. The owner may want to sell the property but cannot. It may be in a deteriorating neighborhood, or may have lost its utility value because of the relocation of a main highway or public access. The property may be too small for necessary production expansion or repairs, renovation, or code violation (fire and safety) would cost more than the current value of the buildings.

In addition, a new business venture may be unsuccessful because of poor management. Or because the business has not met with expected success, there is an urgent need for cash to meet pressing obligations.

Manufacturing plants are burned when machinery has become obsolete; when new production methods have made the process or end product obsolete; when the manufacturer has moved to another location and has been unable to sell or lease the old plant.

In addition, the owner may be having marital problems or need the money to pay hospital bills, medical bills, or attorney's fees. Occasionally, the owner wants to retire or move, but all his capital is tied up in the property, which cannot be easily sold.

Some of these same motives are used to justify the burning of dwellings, apartment buildings, or other inhabited structures. In some larger cities, buildings are torched as soon as all salvageable material has been removed or stolen. In other cases, the building has been sold, but its removal is a condition of the sale.

Sometimes the person whose property is burned is an innocent victim; it is someone else who will gain financially from the fire. It might be a competitor, who wants competition removed. Or it might be another property owner, who wants to expand by buying up the land the building is situated on. Or it might be persons desiring to sell more fire insurance to other property owners in the neighborhood; a contractor who wants to secure a contract to rebuild, wreck, or salvage the building and materials; or an adjuster who wants more business.

In some areas, welfare recipients whose current housing has been damaged by fire are given cash moving allowances and housed in better quarters than they were occupying. This encourages fires in rundown buildings. Other economic motives are the avoidance of bankruptcy, the protection of credit ratings, termination of a lease, or the loss of access to the property.

Another economic motive is intimidation—to force a business to buy protection or insurance, or to purchase merchandise that has been stolen or hijacked. This form of intimidation has been used by both sides in strikes and labor disputes, by criminals in connection with the intimidation of witnesses and for extortion, and by crusaders and terrorists.

It is quite obvious that almost any economic excuse can be and has been used as the motive for the fraud fire. But the investigator must look beyond the obvious. Who else besides the insured (owner) stands to profit from a fire?

Incendiarism in Industry

Industry has become a frequent target for incendiary fires. A study of this problem[4] has developed a number of pertinent points. This type of fire can occur in almost any industrial plant or facility. Its most common location is in warehouse or storage buildings, where fewer people are present and the fire is likely to go undetected for the longest period of time. In the study, it was found that 70% of the fires occurred in storage occupancies. The study further pointed out that of those researched, the greatest number of fires were started by unknown persons. The most numerous group of known firesetters was employees, followed by specific intruders and juveniles. The motives, for those fires in which motives could be established, were fraud, grievances against plant management or other employees, the covering of other crimes, thrills, revenge, maliciousness, and vandalism.

The occupancies most vulnerable to these types of fires are those open to or serving the public, and those where combustible storage is common. Besides the warehouse and storage areas of industrial plants likely targets of these firesetters are schools, hospitals, mercantile property, and shopping centers. The study reported that "simultaneous multiple fires were commonly encountered,"

---

[4]C. W. Conaway, "Incendiary Fires in Industrial Occupancies," *Fire Journal*, 70, No. 2 (March 1976), 28–33.

and that "when an incendiary fire has occurred, there is a strong chance that another will follow at the same plant."[5]

Revenge and Spite Fires

Most malicious fires are the acts of individuals. Revenge fires fall into the category of maliciousness. Lewis and Yarnell found that, in general, there were three categories of revenge firesetters: (1) the offenders who seek revenge upon an employer or former employer, (2) those avenging their own or their families' reputation in the community, and (3) those seeking revenge against institutions in which they were being forcibly confined or that they were forced to attend, or against their own families.[6]

The individual firesetter sets his fire alone and in secret. Lewis and Yarnell found in connection with incendiaries, who are potential arson repeaters, that "revenge is a basic motive behind all firesetting, regardless of age."[7] It must be understood that two motive categories are not being considered in this statement, arson for profit and arson to cover another crime. The latter, however, may be a revenge-motivated crime.

Lewis and Yarnell defined a revenge fire to be any fire that is set directly against an individual establishment toward which the firesetter had knowingly expressed, verbally or through other behavior, a resentment.[8] These included parents, neighbors, schools, employers, and institutions. If the incident that precipitated the act is known or can be identified, the investigator may be aided in determining the perpetrator.

Revenge is a very strong motive for firesetting. The perpetrator of this type of fire does not care about or even realize the tremendous destruction caused by the act. He or she is interested only in "getting even." Many revenge fires are set with flammable liquids under porches, in the crawl space under houses, or in garages.

Spite fires are very similar to revenge fires. They are set to bring discomfort and inconvenience, or to damage or destroy sentimental or personal objects. In one case, a couple and their children had gone to another state to attend a family funeral when the incident took place. First all the food in the refrigerator was thrown on the walls and carpeting. All the husband's clothing was left in the closet but slashed with a knife. Wedding pictures were smashed and covered with gravy and catsup. The planter in the living room was dumped, and all the dirt and plants ground into the carpeting. Only then was the fire started—in the TV room, with gasoline. It was discovered before it had communicated very far

[5]*Ibid.*, p. 31.

[6]Lewis and Yarnell, *Pathological Firesetting*, p. 64.

[7]*Ibid.*, p. 432.

[8]*Ibid.*

and was extinguished promptly. All the evidence of spite and revenge was still very apparent in the kitchen, living room, and dining room and on the second floor.

Other types of spite fires that may occur are the result of neighborhood disputes, racial or religious problems, or family fights, particularly where there is a separation or divorce proceedings. Fires started in beds that are not from careless smoking are likely to be the results of lovers' quarrels. In some rural areas, disagreements have resulted in the burning of homes, barns, and crops. The consumption of alcohol is often associated with this type of fire.

Jealousy- and spite-motivated fires are quite similar. The firesetter may be incited by a wounded vanity and jealous rage, which demand immediate retribution. The desire for reprisal is so great that these firesetters ignore the possibility of injury to themselves. This is a solitary type of firesetting, usually set in secret.

Other types of solitary firesetters include the watchman who wants to be a hero, the teenager who sets the fire in order to help the firefighters, the housewife who wants to keep her husband home nights, the psychotic firesetter, and the sexual deviate who utilizes firesetting for sexual gratification.[9]

Arson to Cover Another Crime

Almost any crime against property can be concealed by a fire. The destructiveness of the fire may assure the perpetrator that no trace of the original crime will remain. Burglary is the most common felony to be hidden in this manner. The offense usually takes place at night in a place of habitation or of business. In connection with burglaries, the investigator should be told about signs of forcible entry found by firefighters. Clear, broken window or door glass from an otherwise smoked-up pane, as seen in Figures 5-12 and 5-13, is another important clue. Arson can also be used to conceal such other property crimes as vandalism, robbery, embezzlement, inventory shortage, and tax fraud.

In the category of crimes against persons, a homicide can be hidden by a fire. Not only may the body be almost destroyed, but many local authorities prefer to think of the fatal fire as an accident. Often, only the autopsy and postmortem investigation will reveal the homicide. In one industrial fire, the body had been interred about a week prior to the arrival of the investigator. Fortunately, the corpse was so badly burned that the mortician had not embalmed it. At the insistence of the investigator, a court order was obtained to exhume the body, and an autopsy was performed. No carbon monoxide above the background level was found in the blood taken from the heart. An intensive investigation followed. It revealed that the man had been the victim of horseplay, and

[9] Bernard Levin, "Psychological Characteristics of Firesetters," *Fire Journal*, 70, No. 2 (March 1976), 37.

the fire had been deliberately set to cover his death. The perpetrators were eventually convicted of manslaughter. Had it not been for the insistence of the investigator that an autopsy was necessary, the crime would have gone undetected.

On other occasions, the victim may not be dead, but unconscious, and the fire used to cover the assault. Weston and Wells observed:

> In arson cases the victim or suspect or both may sustain fire injuries. Such injuries should be recorded by photography, if possible, and verified by a medical examination and diagnosis. Because the weapon is the fire, data on how the fire was started are important, and if possible, the investigator should ascertain whether the injuries resulted from the use of a particular fire accelerant.[10]

A fire has been used to distract public-safety personnel while a major crime, such as a bank robbery, was being committed. Arson has also been used in covering an attempted jail break, or escape from a hospital or similar institution.

Because of the possibility of other crimes against property, the investigator should view with suspicion a fire's timing or coincidence. A fire within hours after the completion of an annual inventory, for example, may be a method of covering up a shortage in the inventory. In one case, the annual inventory of a national chain-operated paint store was completed at about 2:30 A.M. Everyone involved with the inventory had left by 3:00 A.M. At about 5:00 A.M., the paint store erupted with fire. Paint cans were blown across the 75-foot street, breaking the windows of the bakery on the opposite side. The fire destroyed all the businesses in the block, with the exception of the four-story telephone exchange. No investigation was conducted, since local authorities felt that the fire had to have been accidental.

Other conditions or actions that should cause the investigator to become suspicious are fast-burning or fast-spreading fires occurring shortly after the departure of the occupants, or fires in which are found damaged fuel lines or damaged electrical wiring, with evidence indicating that the damage was done prior to the fire. Also grounds for suspicion are electrical appliances—such as hot plates, crock pots, or electrical frying pans—found in an unusual location, which happens to be the area of the fire's origin.

Another crime that fire is quite often used to conceal is embezzlement. A fire's destroying the books just before an audit is a suspicious coincidence. The investigator may find the origin of the fire to be in the office where the books have been conveniently left open on the desk. Or firefighters may discover the books set up with their bindings up and the pages spread out to assist in their destruction by the fire. If the books are not too badly burned, they may be salvaged and still used in an audit.

[10] Paul B. Weston and Kenneth M. Wells, *Criminal Investigation: Basic Perspectives*, 2nd ed. (Englewood Cliffs, N.J.: Prentice-Hall, 1974), p. 266.

Incendiary fires often occur on holidays, during electrical storms, and prior to or during renovation or redecorating.

Firesetting for Thrills

Thrill fires apparently fall into three general categories: those set as an act of vandalism, those set for excitement, and those set for a sexual thrill or satisfaction. Thrill fires set for malicious purposes or vandalism can be set by individuals or by groups. According to Levin, fires normally set by groups or by individuals in the presence of peers include vandalism fires. The presence of others from the peer group encourages the act of firesetting.[11] Some group firesetting has occurred as an initiation into a peer group. For several years, the initiation into a high school group was the burning of barns in an adjoining county. Once the motive for the fires was discovered and the group broken up, the number of barn fires decreased spectacularly.

Riot fires and political fires are also set in the presence of peer groups. The political fire is usually premeditated and set to dramatize a cause.

The firesetter looking for excitement may be interested in the response of the fire apparatus, the excitement of the crowd of spectators, and traffic. He may be setting fires because of boredom. The motives of those setting fires for excitement are closely related to those of the pathological firesetter and will be further discussed with that motive category.

Stekel noted three types of firesetters whose acts have sexual roots and who may be found among pyromaniacs. The first is sexually excited by watching the fire. The second uses fire as a defense, setting fires during periods of enforced sexual abstinence. The third uses firesetting as a total sexual substitute, to free the firesetter from an undesirable sexual habit. Often after the fire, Stekel reported, the habit is checked.[12]

Juvenile Firesetting

Firesetting by juveniles under the age of 12 generally occurs during the daylight hours. At night, most children of this age are in bed or at home—with the exception of those who are not supervised and those, like newspaper carriers, who work during the early morning hours.

Most people are fascinated by fire, men more often than women. Children, being very curious, are no exception. Firesetting from the age of 4 to 7 is mostly from curiosity, or to attract attention. The child plays with matches and lights small fires, which occasionally get out of control.

[11] Levin, "Psychological Characteristics," pp. 36–41.

[12] Wilhelm Stekel, "Pyromania," *Peculiarities of Behavior*, trans. James Van Teslaar (New York: Liveright Publishing Corporation, 1924; Library Edition, 1943), Vol. II, Chap. XI, 159.

At this age, the child may be of either sex. After the age of 7, and particularly as an adolescent, the juvenile firesetter will most likely be male. In their study of 238 juvenile firesetters under the age of 16, Lewis and Yarnell found that 220 were boys and 18 were girls. Among adults, they were able to find only 201 cases of female firesetters, as compared with 1,145 males.[13]

Levin says, "No discussion of the psychological characteristics of arsonists would be complete without mention of the large percentage of arsonists who had childhood enuresis; that is, they were bedwetters as children."[14]

According to Sutton, in her discussion of the juvenile fire problem:

> Firesetting may be classified according to two types: the preadolescent type and the adolescent type. In the preadolescent child the firesetting is primarily an act of aggression or revenge, and it is carried out usually very close to home. In the adolescent it usually has a sexual meaning.[15]

Lewis and Yarnell reported that in adolescent behavior, "the greatest incident of firesetting occurs between the ages of 13–18 years, and the firesetting is associated with the sexual conflicts arising from that period of emotional stress."[16] In addition, they said, "Before 13 years, boys make primarily vengeance fires directed against the home or the school."[17]

Characteristics of the fires set by children are such things as poorly started fires, with evidence of candle drippings and match stems around the point of origin. Usually these children find it necessary to make several attempts before having success. Fires are set in vacant buildings, play areas, garages, schools, and on and under porches.

The typical male firesetter between the ages of 7 and 13 sets his fire for revenge. It may be to get even for either an actual or a fancied wrong—but one that, to him, is real. As an example, a 10-year-old was forcibly ejected from a public building after being caught sneaking a smoke in one of the washrooms by the watchman. Shortly after, he sneaked back into the building and set fire to a waste container in another lavatory. This fire he "discovered" and called to the attention of the watchman, who quickly extinguished it. The boy's desire for revenge was not quite satisfied by this fire, for it had not caused the excitement he had anticipated. So he started down an alley next to a store and, coming upon a trashcan filled with combustible materials, set fire to it. Again he played the "hero," this time by calling the store owner's attention to the fire. But he was still unsatisfied. He returned to the public building that was the scene of his

[13] Lewis and Yarnell, *Pathological Firesetting*, p. 29.

[14] Levin, "Psychological Characteristics," p. 40.

[15] Dr. Beverly Sutton, "Pyromania and Psychopathic Firesetters," *Fire and Arson Investigator*, 25, No. 2 (October–December 1974), 24.

[16] Lewis and Yarnell, *Pathological Firesetting*, p. 311.

[17] *Ibid.*

earlier fire. This time he lighted some papers next to other combustible materials. The resultant fire destroyed the building, with a loss of over $1 million.

In another case, an 8-year-old boy was attacked in a friend's backyard by a large German shepherd dog and two standard poodles that were running loose. The shepherd was chewing his shoulder when the mother of his friend beat the dogs off with a baseball bat. Several days later, the boy set fire to and burned down the garage belonging to the owner of the dogs. This fire apparently did not satisfy his desire to get even, so he set fire to the back porch of the same house. The fire was extinguished by the fire department. His desire for revenge satisfied, he set no other fires.

Chapter 8 related the story of the 14-year-old who set fire to his neighbor's garage when he was refused permission to ride the neighbor's motorcycle. That fire resulted in the deaths of two small children.

Preadolescent firesetting among females does occur. These fires usually involve the girl's personal clothing—perhaps a closet fire or a bedroom fire. One of the more common motives for this type of fire is that the child has never had any new clothing of her own. She has received everything she wears from siblings or relatives, or other secondhand sources. Suddenly, a fire occurs, destroying what clothing she has. Now, the replacement clothing is new—the first new clothing she has ever had, all her own! This motive has resulted in many closet or bedroom fires. Closet fires are also set by males for the same motive.

A male juvenile may set a fire and then masturbate in front of it. He usually leaves the fire burning, since his interest in it has been satisfied.

The burning of twelve to fifteen barns in a rural township was traced to the sexual desires of a female juvenile adolescent. Her "thrill" was to watch the burning barn during intercourse. Her male companions obliged her by setting the barn fires and positioning the car for her best view.

Pathological and Psychopathic Firesetters

Arson of all types requires little physical strength or courage. The crime is a cowardly act. Impulsive firesetters rarely give any thought to the possible injuries or deaths they may cause. They rarely consider the possibility that their fires may communicate to other structures. When they set a fire, they use a match or lighter. Lewis and Yarnell found:

> As a whole, the group coming under the general classification of "pyromania" are not interested in destroying property, but under the impulse merely to apply a match, to the first inflammable [sic] object they find. This may be a baby carriage, a trash pile in the street, furniture, journalbox waste, anything that comes under their eye at the moment, and they usually watch the flame long enough to be sure it gives promise of burning. From that moment on, their attention centers on the firefighters and

the community turmoil that ensues. Accordingly, the fire may be quickly extinguished or assume devastating proportions, depending on existing conditions.[18]

THE WOULD-BE HERO

There are two types of these, the ones who "discover" the fires and the ones who "assist" the firefighters. The first are usually exhibitionists and pathological liars, but they are often well thought of by friends and employers. Their families and friends quickly rally to their defense, frequently assuming the responsibility for their future behavior. For this reason, they are often difficult to convict. They are motivated primarily by vanity. Their actions are impulsive. They may also be capable of other acts such as assault, rape, and theft. This "irresistible impulsive" can take a sudden turn from firesetting to one of these other forms of antisocial activity.

Their greatest desire is to turn in the alarm of the fire, which they describe as being an "irresistible impulse." They do not identify themselves with the firefighting, but rather with the discovery of the fire. Since they desire a blaze worth reporting, their fires may be dangerous in size and may imperil human lives. They find old buildings, sheds, buildings, or homes under construction, any type of structure that will take fire rapidly and create a large, spectacular blaze. These people almost scream for attention by their actions.

The adults in this group may be alcoholics; the juveniles simply have the desire to play heroic games. They are interested in a large fire with lots of flames and smoke, since this is the type of blaze that attracts many spectators. As the one to have discovered the fire, such a person basks in the limelight. They themselves and the spectators now consider them to be the "heroes."

The second type of would-be hero desires to identify and be identified with the firemen—with the powers that extinguish the blaze. This person is not interested in the fire's destructiveness, as is the first type and the revenge firesetter. This is the would-be fireman. He helps pull off the hose and raise the ladders, and generally gets in the way of the fire-suppression activities. He may also perform rescues, all to demonstrate his masculine ability to overcome this destructive agent—fire. As a preadolescent, he was the type of boy who took great pleasure in setting fires and putting them out by urinating. Now he assists in extinguishing the fire by using the fire hose. Many of this type associate the fire nozzle and fire hose with a giant penis, and the water with urine.

Another "hero" type of firesetter is the person who feels insecure. He sets the fire to reinforce his job as a watchman, guard, or industrial or institutional firefighter. If he has no "action," no fires to discover or put out, his job may be abolished. In this category are the firesetters who are trying to establish the need for a watchman, guard, or firefighter. It would be convenient for the plant to

---

[18] Lewis and Yarnell, *Pathological Firesetting*, p. 33.

which they are applying for a job to have a fire. Then the need for their services will be forcibly demonstrated.

Finally, a mentally defective or physically handicapped person may set a fire, then turn in the alarm, thereby "proving" to the family and the community that he is alert and capable.

THE PYROMANIAC

These are the mysterious "firebugs," whose activities terrorize neighborhoods and communities. They usually set a series of fires in rapid succession at night. They set trash fires, and when these do not satisfy their desires, they move to other combustibles. There is no reason or material profit in the act.

These people usually complain of mounting tensions, restlessness, the irresistible impulse or desire to set a fire. They have headaches, palpitations, and ringing in their ears and are unable to control themselves in the act of setting the fire. The firesetting is not planned. There is no detailed preparation, and flammable liquids are not procured for the act. The material burned or used to set the fire, including any flammable liquid, is whatever may be available at the scene. Favorite places are in hallways behind a stairway, in baby buggies found in a hall, on stairs—for that matter, any area that will burn fast and create excitement. Once the fire is started and the people are in a state of panic, once the fire department responds, the tension passes. The firesetter loses all interest in the fire and probably goes home to sleep.[19] The *modus operandi* (method of operation) or pattern of this type of firesetter, once established, is usually repeated.

Each individual pyromaniac seems to develop his or her own special pattern. Fires may be set on a certain day of the week, or at a certain time of the month—for example, when the moon is full. Or they may select a certain time of day. The types of occupancies and buildings the pyromaniac seems to choose are churches, schools, multiple-dwelling occupancies, lumberyards, specific types of homes, certain offices, or garages. The pyromaniac is quite likely to attempt to set a series of fires, most frequently a series of three. They may not all be successful, but they often are. These fires may be set on the same night, or they may be set a certain number of nights apart. There will, however, be a consistent pattern.

In one case, the arsonist set fire to unlocked garages, all in alleys. The pattern established was quite precise. The fires were all set on the same night of the week, between 1:00 and 4:00 in the morning. In all instances, a fire was set in a garage in one block and a second fire in another garage in the adjoining block. Typically, the arsonist brought no devices or material with him. He used whatever was available in or about the garage. The fire department would be fighting one fire when the second was started. At one fire, he asked one of the firefighters for a match to light a cigarette. The firefighter handed him a pack

[19] Lewis and Yarnell, *Pathological Firesetting*, p. 87.

of matches and said, "Keep them." He did, and proceeded to the next block and used the matches he had just borrowed to set the second garage on fire. He was finally caught after stakeouts were established.

THE VOLUNTEER FIREFIGHTER

Lewis and Yarnell report over 150 cases of arson involving volunteer fire-fighters, "fire buffs," or men who tried but failed to become firefighters.[20] These men set fires, sometimes turn in alarms, but always are available to fight the fire and assist in extinguishing it. This group is closely related to the would-be hero. However, in addition to being the hero who "discovered" the fire, the hero who performed the rescue, other motives are responsible for the firesetting activities of these men. Many are paid-on-call—that is, paid for each fire call to which they respond; the greater the number of fire calls, the greater is his income. This, alone, could be a very strong motive. Others want to be able to operate a new piece of apparatus, or set fires in the district of another fire company to injure its reputation. Alcohol is quite often a factor in the behavior patterns of this group. Sometimes the motive is no more than the desire to create some excitement; others set fire to vacant structures in order to "get some practice."

ADULT FEMALE FIRESETTERS

Most generally, women set fires for revenge or to attract attention. As we have seen, there are fewer female than male firesetters. A fire set by a woman is usually within her own personal sphere of interest—in her home, her neighborhood, her church, her place of work, somewhere within the area of her personal domain. The fires are usually small and many times will burn out by themselves. In churches, fires of this type are set among the choir robes or near the organ.

Those fires set for revenge are in the residence or business of the one the firesetter wishes to injure. Fires in or on a bed or in the closet to destroy clothing are typical. Jealousy and revenge concerning the "eternal triangle" influence this firesetter. Separation and divorce proceedings can precipitate the action. In other cases, a woman may be trying to get even with her parents, who restrict her social activities even though she considers herself to be an adult. Or she may be jealous of a sibling, whom she believes to have a successful and happy marriage in contrast to her own.

Those fires set for the purpose of attracting attention are also within the woman's personal world. These are usually confined to her own home. It may be that she wants more attention from her husband or boyfriend. She may have mailed threatening letters to herself, or reported receiving threatening phone calls or being subjected to Peeping Toms or other traumas, in hopes that the errant husband or boyfriend will stay home. Then, finally, she is the victim of a fire—not a serious fire—and has to be rescued.

[20] *Ibid.*, pp. 193–227.

Summoning the fire department may become a motive for firesetting when a female firesetter is interested in one of the firefighters. The fire will occur on the shift he is working and in his district, so that he can "rescue" her from the fire!

Another type of female firesetting occurs within the period just prior to or during menopause, when emotional upheavals occur.

Sabotage, Radicals, and Terrorism

Many historians and other authors have pointed out that ours is a nation conceived in, born of, and developed by violence. The Sons of Liberty and other patriotic groups of the American Revolution are examples. Many groups have used violence to further their causes in the history of this country. Fire has always been a tool of radicals and terrorists. The use of arson during the Civil War by military and civilian groups is well documented. Sabotage by fire and bombings during World War I, World War II, the Korean War, and the Vietnam conflict was not uncommon. Nor was the sabotaging of fire extinguishers, as for instance in the substitution of fuel oil for water in pump cans and pressurized water extinguishers at military munition plants during the Korean War.

Saboteurs and terrorists work in basically the same way. They are all goal-directed, having a purpose or cause. They differ only in that saboteurs discriminate in directing their activities toward a particular group or person. Terrorists do not. Their target group is universal; their desire is to intimidate and create fear on the part of the victims or potential victims. No group is exempt from attack. This unpredictability generates insecurity and loss of confidence in government agencies and their ability to protect the citizen from attack. The use of fire bombs of various types is common. Nothing strikes terror in the heart of the victim as much as fire.

QUESTIONS

1. What is the principal study on pathological firesetting?
2. List at least eight motives for arson other than the profit motive.
3. You are security chief of an industrial complex; what areas of your complex lend themselves to firesetting activities?
4. What are the more common motives for arson in the industrial setting?
5. In their study of pathological firesetters, what were the three general categories for revenge-type fires found by Lewis and Yarnell?
6. What are the two most common motives for firesetting today?
7. Describe the three types of firesetters Stekel noted as having sexual roots to their firesetting motives.

8. Dr. Sutton classifies juvenile firesetters according to two types. What are they?

9. What does Dr. Sutton believe are the primary motives of each of these classifications?

10. Discuss the two types of "would-be-hero" firesetters. Explain the difference between the two.

11. A series of fires develops a special pattern as to time of day, day of week, time of month, and types of occupancy, and the fires are set in the same manner. What type of firesetter would you suspect?

12. What are the principal motives women have in setting fires?

13. Toward what group does the terrorist target his firesetting activities?

11 Interviews and Interrogations

Skill at interrogation is not some elusive ability found only in the master investigator. It is simply the ability to interview a suspect or witness and obtain the required information from that person. The arts and skills that distinguish the good interrogator from the great interrogator can be developed. They consist mainly of an understanding of the human mind and how it functions when being questioned. This chapter will discuss some of the fundamentals of the successful interview. The method described should dispel many of the groundless fears faced by the potential interviewer.

Two important U.S. Supreme Court decisions define what the Court considered to be improper and illegal interrogational procedures. The *Escobedo* and *Miranda* decisions limited the methods that may be used by the investigator in the interrogation of a suspect. As we saw in Chapter 3, the Supreme Court defined the permissible types of searches and indicated when a search of the fire scene became a criminal investigation. So, too, it has drawn a line between the interviewing of a witness and the interrogation of an individual when that individual becomes a suspect.

So important are the rights of the accused that the interrogator (investigator) must be able to properly communicate them to the accused. The person being interrogated must fully comprehend these rights. Failure of the suspect to understand what has been said can result in the accused's release by the court. Any evidence or information acquired in such circumstances will be illegal, and the results to the case can be disastrous.

The *Escobedo* and *Miranda* Warnings

The first and probably the greatest block faced by the interrogator will be the *Escobedo* warnings:

1. You have the right to remain silent.
2. If you give up that right, anything you say can and will be used against you in a court of law.
3. You have the right to talk to a lawyer and have him present with you while you are being questioned.
4. If you cannot afford to hire a lawyer, one will be appointed for you, without any cost to you, prior to questioning.

After reciting these to the suspect, the investigator must ask the following questions, derived from the *Miranda* decision:

Do you understand each of these rights I have explained to you? If you give up these rights and agree to talk to me without a lawyer present, you have the right to stop at any time you wish. You may also demand a lawyer at any time.
Having these rights in mind, do you wish to talk to me now?

These warnings must be given only by law-enforcement officers, or by persons acting at the direction of, or as agents of, a law-enforcement officer. The rights apply only to a person being questioned; they do not apply to free and voluntary statements made without questioning. They also apply to a person being questioned in a hostile environment, such as a police station, or to a subject who has been placed under arrest or is a suspect in a case.

Let us examine each warning and explore its meanings. When the warnings should be given and when to avoid giving them are matters that have been discussed in many law treatises. The following material may be helpful in the development of a personal style of interrogation.

The first two of the *Escobedo* warnings are, "You have the right to remain silent. If you give up that right, anything you say can and will be used against you in a court of law." These have been interpreted by the courts to cover both inculpatory and exculpatory statements. An inculpatory statement is an incriminating statement. An exculpatory statement is one designed to prove innocence or to vindicate a person. This means that statements of both guilt and innocence are covered by *Escobedo* and *Miranda*. One might question why a defense attorney would try to have his client's statement of innocence thrown out of court. If the client has lied repeatedly in statements of innocence and the investigator has been able to prove these statements false, the image of the client will be injured in court, since he or she will be shown to be a liar.

The third warning is, "You have the right to talk to a lawyer and have him present with you while you are being questioned." If the suspect wants to call a lawyer, or the lawyer states a desire to be present when anyone talks to the client, the lawyer must be notified. If the call to the lawyer is not permitted, anything the suspect says will not be allowed in court. If the subject can be persuaded to waive this right, he must be convinced that it is in his best interest to confide in the investigator. However, the investigator must be extremely careful that the persuasion cannot be construed as coercion.

Fourth is, "If you cannot afford to hire a lawyer, one will be appointed for you, without any cost to you, prior to questioning." This is a protection for the poor and refers to the services of the public defender's office. Generally, these attorneys are appointed by the court. If the suspect requests an attorney or a public defender, the interview is over until the attorney arrives.

Next, the interrogator must ask the questions directed by *Miranda*, as mentioned on p. 144, in order to ensure that the subject understands all the rights to which he or she is entitled.

When are these warnings applied, and when can they be avoided? In an initial interview that is conducted at the time of the fire, at or near the scene, the warnings can be omitted. But if you have strong evidence that the subject may have committed the crime, the warnings must be given.

As an example: A house has burned down. The owner is being interviewed in a car about 1,000 feet from the fire scene. At this time, the questions being asked are those necessary for the record: his name, the date of his birth, where he works, the name of his insurance company, how much coverage he has, who his agent is, what time he left the house, where he has been. All this information could later be damaging to him through court proceedings. However, at the time they are asked, he is not a suspect. The investigator is simply gathering information needed to begin the investigation. But as the subject is being questioned, the odor of gasoline is noted about his person. If, at that point, it is felt that he may be involved in the setting of the fire, he becomes a suspect. When the focus of the investigation begins to point toward him, then it is time for the *Escobedo/Miranda* warnings and restraint of his person. What does "restraint" mean? He is asked about the smell of gasoline. If he does not want to discuss it and states that he wishes to get out of the car, he is told to stay there and given the warnings, and the questioning is continued.

Another situation might be the visit of an investigator to a person's home or place of business. The investigator introduces him- or herself and asks to speak with the person regarding the case. If the person states he has no objection, the investigator may proceed with the questioning. The person being questioned is not in a hostile environment, nor are his actions being restrained by the investigator. Therefore, the warnings can be omitted. However, if the person says he does not want to be questioned, the investigator may suggest that any further discussion may take place either where they are or "downtown." At this point, the entire nature of the interview has changed and the warnings should be given. Remember, the investigator sets the mood of the interview. In many cases, the need for the warning can be prevented and a successful interview achieved.

The *Escobedo/Miranda* warnings do not have to be given under certain conditions. These are:

If the investigator is not a law-enforcement officer.

If the investigator is not interviewing the subject because a law-enforcement officer has directed him to do so

If the investigator is not an agent of a law-enforcement officer.

NECESSITY OF THE WARNINGS

The warnings are necessary. The fire investigator must be certain that these rights are, in fact, granted to the subject to the full degree guaranteed by law. The warnings apply to statements of both guilt and innocence. Therefore, it is best to read them to the person when they are given. If the printed form is used, have the suspect sign the form. Be sure that the person signing can read and write. One way is to ask the suspect to read the warnings aloud as the form is signed. If the suspect cannot read or write, he makes his mark which must be witnessed by two people. In order to diminish the effect of this, explain that the signing of the statement is a required part of the interview. Do not continue to repeat the warnings. And once a waiver statement has been signed, *do not lose it.* A lost waiver statement can destroy the best possible case.

It is generally understood that people who are witnesses rather than suspects in the investigation do not need to be warned. Care must be taken by the investigator, however, that any statement a witness makes does not involve him or her as a suspect. Any statement made before the warnings were given will be carefully examined by the court. The statement must be that of a witness and not of a suspect. A ruling by the court that the witness's rights were violated by the statement may destroy the case.

Physical Setting for the Interview

A great deal of planning and care is necessary to obtain the maximum amount of information from the interview. The place selected and its surroundings are extremely important. The room chosen should be free of distractions, so that the subject's full attention is focused on the interviewer. There should be no telephone in the room, no charts or pictures on the wall, no items of interest present on the table or desk. If possible, the room should be windowless. If there are windows, drapes should be drawn. Ideally, the drapes should be the same color as the walls of the room. The color of the room is of great importance. It has been found that strong greens are most conducive to conversation. Avoid any aggressive colors, such as red or orange, and relaxing colors such as pastels or earth tones. The interviewer should be the only object in the room upon which the subject can focus his attention.

A bare desk or table upon which the interviewer can place a notebook is necessary. Two chairs should be provided, both of the same general material. However, the chair of the interviewer should have arms and should be slightly higher than the other chair. This tends to create the illusion of dominance by the person conducting the interview. It may be very important that the investigator not attempt to write any notes during the interview. Note-taking has resulted in stopping many a confession.

The interview of a witness is a face-to-face conversation to obtain information about the fire. If the person being questioned is under any restraint or is suspect, the interview becomes an interrogation. The interviewer must be very careful not to lead the witness. Questions that appear to require a given answer may lead the witness to supply information suggested to him, but actually not known to him. This may damage or even destroy the witness's true recall of what he saw.

A tape recorder is a necessary tool. The tape can be replayed at a later time and reproduce every word spoken during the conversation. The recorder should be left in full view of the witness from the beginning to the end of the interview. The investigator should call the witness's attention to it and obtain consent for its use. The tape recorder can also be used in questioning a suspect—again, after obtaining consent for its use. In both cases, however, the result of the conversation may be less then ideal. This result is due to the inhibitions that the use of a recorder instills in people. The primary advantage of the proper use of recording devices is in obtaining far more information, with a greater chance of discovering falsehoods and discrepancies.

One of the more effective procedures is to conduct the initial interview without any recording device and to refrain from taking any notes. Follow this first conversation with a request that the second (follow-up) interview be recorded. The reason for the recording can be that it will permit the secretary to make an accurate transcription of what has been said, to minimize the possibility of misquoting the person being interviewed. Explain to the witness that the transcription will be available to read and sign. Most of the time, once a witness has discussed the fire with the investigator, there are no objections to repeating it at a second session.

The second interview, which should take place immediately following the initial one, serves a twofold purpose. First, any forgotten information may come to light during the second interview; and second, falsehoods, which are hard to maintain, will surface.

Another important point: Be sure that several obviously misspelled words are included on each page of the transcribed statement. Have the person being interviewed discover and correct the misspellings, in ink. Both the investigator and the witness should then initial each correction. This shows that the person actually read the statement. It eliminates a later claim that the statement was signed without being read.

There are two different approaches to the number of people that should be present at the interview. One holds that only the witness and the interviewer should be present. This may be very important with a witness, and sometimes imperative with a suspect. In order to reach the desired successful conclusion, the interviewer must develop a rapport with the person being interviewed. The investigator must be looked upon as a friend. This may be difficult to achieve when another person is present. The presence of a third person may actually inhibit the witness to the point of not being responsive.

The establishing of rapport between the witness and the investigator permits the witness to confide and relate incidents that otherwise would remain untold. This rapport takes the form of conversation that breaks down the natural reluctance to give information to authority figures. Often, witnesses fail to give information simply because it was not asked of them. When the witness and interviewer develop the ability to talk to one another, the result is far more rewarding. And what is helpful in questioning the witness is a matter of necessity with a suspect.

Under other circumstances, the interviewer may desire a third party present as an official witness during the proceedings. This situation may occur when the person being questioned is a female and the investigator is male. It must be thoroughly understood that the third person is an official witness to the proceedings *and no more*. He or she is present but *does not participate* in the questioning. One and only one person does the questioning. Silence on the part of the third person is most important. After the interviewer has completed the questioning, the third party may be asked if he or she has any further questions. Then, and only then, should the silence be broken. A third participant to the proceedings who continually breaks silence can destroy the entire interview. For this reason, many investigators prefer to conduct interviews alone.

The days of beating or forcing a confession from a suspect are gone. A habitual offender knows the old "good guy–bad guy" routine as well as the people using it do. The first offender is usually scared by it and does not hesitate to relate his fear in court. This will throw a cloud upon confessions or statements and all information gained from them. It may result in the exclusion of the confession and much of the evidence. No matter how repulsive the crime, no matter how repulsive the offender, the suspect must be convinced that the interrogator is a friend.

But this assumption of friendship does not mean that the suspect is an equal. The investigator must always maintain a position of dominance—although in a subtle manner. The slight difference in the height of the chairs and the arms on the interviewer's chair aid in this. The interrogator should direct the witness/ suspect to sit in the chair intended for the person being questioned. The interviewer must maintain control of the interview at all times, asking the questions, not answering them. If the suspect begins to ask questions, put him off politely or just ignore the questions.

Some witnesses are eager to tell what they saw and what they perceive to have happened. Others, more reluctant—as well as suspects—may have to be led into such a discussion. The investigator may find it necessary to ask about such things as the person's hobbies and family to get a conversation started. The interrogator should never begin the questioning of a suspect by asking him to confess or by direct questions about the fire. It is important that background information about the suspect be obtained—such things as his interests, hobbies, any type of conversation to get him talking. Once the suspect begins to talk, he may continue to do so; do not attempt to stop him. Again, the suspect must feel that the person interrogating him really cares about him and his problems. He must believe that he has really found a friend. Once the suspect has been convinced of this, it will not be too difficult to persuade him that to please this "friend," it will be necessary to confide in the "friend."

The interrogator must have an understanding of the human mind. Sometimes the simplest apparently insignificant gesture, action, or word will ensure success. The next time, however, that same gesture, action, or word might fail. If the interrogator indicates that he does not believe the suspect, that he thinks the suspect is being dishonest with his "friend," the suspect may begin to tell the truth to gain acceptance and forgiveness. A very important factor is establishment of some type of physical contact. Different individuals and different sexes will require different approaches. The contact can be as simple as a pat on the hand, or an arm around the shoulders. Both can indicate sympathy for the suspect's problem. This gesture must be carefully planned so as to appear genuine and spontaneous. The interrogator should never lose sight of the goal.

Another method of securing information or a confession is by playing upon the person's weaknesses, including pride and vanity. Such questions as what all this must be doing to his family, what his friends will think of his actions, or what effect the situation will have on his job might be asked. The interrogator must have some knowledge of the suspect. The effect his situation will have upon his standing in the community or in his church will be a strong influence upon one person and mean nothing to another. The basic premise is that the results of his actions must be dire. But by confiding in his "friend" he has the opportunity to neutralize what has happened. He must be convinced that the interrogator is the only one in the world who can and will be able to help diminish the consequences.

When questioning a suspect, never try to bluff him unless you can carry it off or back it up. Do not tell him that he was seen at the scene, or that his fingerprints were found there. He may call your bluff and tell you to charge him, that he is through talking and wants a lawyer. The bluff, if necessary, should be vague and take the form of a question. He might be asked if he realizes the effect of having his prints found at the scene. Or, he might be told that a house-to-house canvass of the area will be made to see if anyone can identify him or place him at the scene.

If you want to know how the suspect committed the offense, do not tell the suspect your pet theory on how it was done. If you are wrong, you will be playing into the suspect's hands. Instead, play on the suspect's ego, hinting at what has been established by the scene search. Let the suspect fill in the gaps and details. Sometimes the approach will be that you did not think the suspect had the intelligence to carry off as skillful a crime as was committed. Now, however, you have a great deal of respect for his abilities. Praise his ability and probe his knowledge in the area of the offense. Careful questioning can elicit such information as:

Does the suspect have a background in explosives, or some knowledge of them?

Was some form of electronics involved?

What is the suspect's knowledge in the field of electronics?

The directions of questions are numerous and should be developed by the interrogator.

The interrogator must be cautious in using slang or technical terms, even though the suspect uses them. Whether you know them or not, encourage the suspect to explain them to you. It is another way to get the suspect to talk. Use everyday langauge and terms. By misusing slang or street language, the investigator permits the suspect to get the upper hand. He may not show it, but he will be laughing at you, for you will have lost control. Do not resort to profanity. If the situation calls for the investigator to lose his temper, he should do it only to further the goals of the investigation. Try to indicate that the problem is not the suspect's failure to confess, but his throwing away the opportunity to let his "friend" help him with the problem he is facing.

Psychologists tell us that most people who commit a crime subconsciously harbor within them the desire to be caught. Many criminals reveal this desire in the method they use to commit the crime. Others reveal it in the way they prepare for the crime. The most obvious revelations are made in the things they say and facts they reveal in their statements.

For example, in an arson-for-profit case, the offender may have made some recent adjustments in the insurance coverage and did not allow enough time to pass to conceal the motive. He failed to wait for two reasons. First, the urgency of the circumstances drove him to the crime. Second, the excitement of getting away with something of a criminal nature spurred him on. When questioned about the increase in insurance coverage, the suspect will begin to supply facts and figures to justify the transactions. He will usually supply too much information, much of which may prove to be false. Talking too much also indicates an attempt at hiding something.

In a recent case, the offender's car was found at the scene of the fire. Since he owned the business, this fact in itself was not particularly damaging. However, on the front seat of the car was an envelope addressed to the insurance company. The envelope was unsealed. Inside were photos of the business's stock on hand, a complete inventory of the stock and fixtures, and accounting sheets showing the amount of business during the last six months. The fire failed to destroy the stock and the records. What remained after the fire proved the records and photos in the car to be false. This established a motive for the crime. Remember, most people are not prepared to handle a natural disaster. Rarely do they have immediately available all the necessary facts and papers to be supplied upon request.

It is important to review all statements obtained and facts gathered. In addition, someone who was not present at the original interview should review the statements. Having been so deeply involved in obtaining the statement, the interrogator may have overlooked an important point.

As indicated previously, the witness is interviewed; the suspect is interrogated. Most of the time, the witness being interviewed is innocent. But if the witness becomes suspect, the *Escobedo/Miranda* warnings must be administered. Keep in mind that there are people who, for various reasons, cannot be interviewed or interrogated successfully. This should not be considered a personal failure on the part of the interrogator. One must learn from failure as well as success.

The interviewer has a very important job to do. With him rests the responsibility to protect and safeguard those citizens who rely on, confide in, and trust him. The interrogator has an equally important job—to prevent the offender from furthering his career in crime.

Taking the Statement

Whenever a witness is interviewed or a suspect interrogated, it is an excellent idea to take a photograph of the person, usually with Polaroid camera. On the back of the photo should be noted the identity of the person photographed, the date, time, and place the photo was taken, the case number, and the photo's relation to the case. In the case of a suspect, it is a good idea to take both before and after photos. Suspects have been known to lie about their treatment while in custody, particularly with reference to the taking of statements. The conditions under which the confession or statement was obtained should be clearly shown in the photograph. This will be valuable in court to back up testimony that the statement was not extracted from the suspect by force.

Photographing a witness will assist in the identification of the witness and refresh the mind of the interviewer. This may be of great importance when the case comes up six months or several years and many cases later. In addition, a witness or suspect may not remember speaking with the investigator or may, for one reason or another, deny ever speaking to him or her. The picture taken at the time of the interview/interrogation can go a long way toward assisting recall.

When making a tape for transcription, establish and follow some type of standard format. The following is suggested as one that might be used: "This will be an interview in regard to the case number _____. This interview is being conducted on [date] at [time]. Present are . . ." At this point, have each witness or suspect speak his or her name, spell it, and give date of birth. Having each person speak will aid the transcriber in the job of identifying the speakers later. Following this, give the location of the interview. Then advise the parties present, identifying each by name, that the conversation is being recorded. At this time, obtain verbal consent on the tape for the recording. If your prosecutor feels it is necessary to incorporate *Escobedo/Miranda* warnings on the tape, this is a good time for it. (This is usually only for a suspect and not needed for a witness.)

At this point, the taped interview/interrogation should begin. Keep the conversation on the tape relevant to the matter under investigation. This is not the time for personal conversation with the person being interviewed/interrogated. During the conversation, if it becomes necessary for any reason to turn off the tape recorder, indicate that you are stopping the recording, the time of the break, and the time when you resume taping. The tape must be retained. It may be necessary to play it in court before the judge and jury. If the time of the break and the time of resuming the recording are not included, the defense can raise several questions: "How long was the break?" "What was happening to the suspect during the long break of no taping?" The original tape must be retained until the prosecuting attorney or the court releases it as evidence.

Tapes can be used to refresh the memories of your witnesses prior to their testimony. If the witness does not recall saying what is on the statement, hearing it on the tape is quite likely to remind him or her.

As soon as possible after the interview, the complete report of the interview should be typed. In typing this report, refrain from using such terms as *interrogation, questioned, confessed, the suspect,* and so on. Instead, use the words *interviewed, discussed, spoke to, he* or *she related, the subject of the interview, the witness,* or the witness's name. This will sound much better in a court presentation, creating a more favorable impression of you and your case in the minds of the judge and jury.

The following is a checklist for the interviewer or interrogator:

1. As the interviewer, am I mentally prepared for this?
   a. Have I read the reports and do I have all the available information at my disposal?
   b. Do I fully understand exactly what information I am looking for?
   c. Do I understand the crime involved, and do I know what elements are needed to prove the *corpus delicti*?
   d. Do I have my emotions under control, and am I prepared to act in a manner that may be necessary to gain the confidence of the subject?
   e. Do I have confidence in my own ability as an interviewer?

f. Am I ready to turn the interview over to someone else if I fail to make any progress with the subject? (Pride—"If I can't do it, no one can"—has caused the failure of many interviews/interrogations.)

2. Am I physically prepared for this interview?
   a. Am I rested and alert, and can I stay this way for more than two hours?
   b. Is my physical appearance good enough for this interview? (Look in a mirror and ask, "Would I talk to the person I see there?")
   c. Am I offensive in any physical way—body odor, breath odor, open sores such as cold sores? (This type of condition can ruin an interview.)
   d. Are my clothes such that they could distract the person being interviewed? too loud or flashy?

3. Is my attitude proper for the interview at hand?
   a. Am I ready to relate to the subject of the interview?
   b. Am I ready to put aside personal feelings about race, religion, etc., for the good of the interview?
   c. Am I ready to be neutral in this matter and keep an open mind?
   d. Can I avoid the pitfall of slanting what is said to what I want to hear?
   e. Am I ready to restrain my temper, even in the face of an obvious lie?
   f. Am I ready to be a good salesman, and sell myself as a friend?
   g. Can I sell my help to the subject?
   h. Will I stay with the interview as long as there is any chance of a successful conclusion?

4. Am I prepared for the person being interviewed?
   a. What is the person's education?
   b. What are his or her likes and dislikes?
   c. What does this person fear?
   d. What is the person's emotional state at this time?
   e. Why is he or she in this condition?
   f. How can this person be expected to react?
   g. What will I have to do to obtain the desired reaction?
   h. Has this person any criminal background?
   i. Have I talked with someone who has interviewed this person before, and if so, what can I do to achieve the proper response?
   j. Will this person lie, and does he or she have the ability to stick with a lie?

5. Has the proper place been selected for the interview?
   a. Does the place selected offer freedom from distraction?
      (1) No phones
      (2) No pictures
      (3) Either no windows or drawn drapes
      (4) No radios, police or otherwise
      (5) Nothing on the desk or table to pick up or play with
      (6) No ashtray in reach of the subject
   b. Is the room well ventilated, with proper heating or cooling?

    c. Is the room private, with a lock on the door?
    d. Is there a Do Not Disturb sign for the door?
    e. Is the room of proper color to achieve the mental state desired?
    f. Are there any weapons or objects that could become weapons in the room?
    g. If there is a telephone, can it be removed or placed out of service?

6. Preparing the subject for the interview:
    a. Are the *Escobedo/Miranda* warnings needed?
    b. If they are needed, have I given them properly?
    c. Is the subject physically prepared?
        (1) Does the person need to use toilet facilities?
        (2) Does he or she need food or drink?
        (3) Is this person taking or in need of medicine?
        (4) When was the last time he or she slept?
        (5) Has the subject been drinking?
        (6) Is he or she an alcoholic?
        (7) Is he or she a diabetic or epileptic?
    d. If the interview is to be tape-recorded, has the subject given permission?
    e. Have I properly identified myself and explained why I am talking with the subject?
    f. Have I asserted my position and achieved the mental dominance that is necessary?

This list is designed to assist the interviewer in preparing for the task at hand. The interviewer must be mentally and physically ready for the interview/interrogation. He must approach it with a positive mental attitude and the desire to be successful. The list covers most of the things over which the interviewer has control.

How can the interviewer prepare for the actual session with the subject and use the proper approach? There is no easy way; practice, experience, and study are the only answers. Studying psychology, even the basics, will aid in the understanding of people. Study body language, something that actually works in practice. It can give you good insight about your subject. In addition, it will allow you to control your actions and not telegraph your moods or thoughts to the person being interviewed/interrogated.

If the investigator cannot understand and reach the person being questioned; failure is likely in some degree. Today's interviewer must stay within the boundaries set by the law. The key to success is the ability of the investigator to mentally prevail over the person being questioned. By the proper use of knowledge, the interviewer/interrogator can obtain the information desired. If one point is to be stressed above all others, it is that *the investigator must utilize his mind to the fullest*—develop it, exercise it, educate it. It is the tool that will be the deciding factor in success or failure.

QUESTIONS

1. What are the *Escobedo/Miranda* warnings?

2. Describe at least five situations in which the interrogator is *not* required to give the *Escobedo/Miranda* warnings.

3. Describe the ideal type of room for conducting an interrogation of a suspect.

4. Why is the "good guy–bad guy" routine unacceptable today?

5. Under what conditions is it advisable to have a third party present during an interrogation?

6. Describe how you would introduce the use of a tape recorder in an interview or interrogation.

7. Discuss some of the advantages of using a tape recorder; the disadvantages.

8. Describe the procedure you would follow if you found it necessary to turn off the tape recorder during the interview.

12 Recording the Investigation

In recent years, there has been a great increase in the number of civil suits arising from all types of fires. These suits have been filed for many different reasons. When the loss has apparently been caused by defective equipment, unapproved equipment, negligence, or faulty installation, people seek relief in the courts. Or after the insurance company pays for the loss as covered by the insurance policy, it files a subrogation suit in the name of the insured. This suit attempts to recover the costs from the responsible party. Or an uninsured victim seeks compensation for personal injuries or for property damage caused by a fire.

This type of civil action is very likely to involve the fire department through fire incident reports, fire inspection reports, or fire investigation reports. The clarity and detail of these reports can have a tremendous impact on the fire department's public image. Reports may be brought into court several months or even years after the fire. In some sections of the country, civil suits are not heard for from seven to eight years after they are filed. So if the reports give only a date, time, and location, and a list of the extinguishing equipment used, they do little for the litigation or for the image of the department. Reports that detail the point of origin and cause, the surrounding physical conditions, and the actions of spectators and occupants create respect for the department. In addition, the injured party may have a just claim and be depending upon the fire department's report to substantiate it.

Methods of Recording the Investigation

Among the most likely methods to be used in recording the fire investigation are notes taken at the fire scene by the investigator, sketches and photographs made at the fire scene, and records made by using a tape recorder at the scene. Other sources of information that may be useful in developing the investigation and completing the report are fire incident reports, fire inspection reports, and financial reports.

Notes should be taken at the fire scene in a convenient-sized notebook. The stenographer's notebook is about as large as is practical. And the notebook should be bound, rather than loose-leaf or the spiral type. Some courts have sustained a motion by the defense to exclude loose-leaf and spiral-type notebooks and bound notebooks with pages torn out.

The investigator records his or her activities and observations in detail. It is not unusual for several months or a year to pass before a case comes to trial. In that time, depending upon the work load, the investigator may have been involved in any number of fire investigations. It is by these notes that the investigator is able to accurately recall the information needed for the report and for testimony in court. Some investigators use both the notebook and the tape recorder for taking notes. Some system of marking and numbering the notebooks should be developed so that they may be filed properly.

All notes should be thorough, accurate, and neat so that they are easy to transcribe and read. The investigator will be referring constantly to these notes during the course of the investigation. They will assist in correlating observations and developing leads. These notes are the basis of all reports the investigator completes.

The Fire Investigation Report

The final written result of taking notes, recording observations, and interviewing firefighters and other people is the fire investigation report.

FIRE RECORDS

Fire records, or fire incident reports as used by fire departments, indicate that some type of incident involving fire has occurred. These records set forth the location, time, date, weather conditions, names of the owner and occupant, a description of the structure, and to some degree, the details of what occurred at the incident.

In addition to the fire incident report, another report may be of interest and of value to the fire investigator—the fire inspection report of the fire-prevention bureau. It should be a fairly comprehensive survey and description of what the premises were like at the time of the last inspection. All violation notices that were issued, corrected or uncorrected, should be recorded in the report. These may be of value in determining existing conditions that could have caused or spread the fire.

The fire incident report should document the need for an investigation by indicating the possibility that the fire had a suspicious, incendiary, or undetermined origin. The fire investigation report will be a more detailed study of what occurred at the scene of the incident. Besides the basic information described above, this report should include the construction and size of the structure, what the firefighters observed and encountered upon arrival, the color of the smoke and flame, and the intensity and location of the fire. A good report will also indicate such things as multiple points of origin and odors that caused the firefighters to suspect the presence of flammable liquids. Ammonia and similar strong odors have been used to cover those of flammable liquids.

DEVELOPING DATA FOR THE REPORT

After the investigator has completed his preliminary survey of the premises, he will begin to speak with people on the scene. The initial conversations or interviewing will usually be with the fire-department personnel. Many of the points the firefighter may observe that are of value to the investigator have been covered in Chapter 4.

What did the firefighters observe prior to and upon arrival at the fire scene? Did they observe anything that could be considered unnatural or not normal under the circumstances? Did they have trouble entering the structure? Was forcible entry necessary? It may be advisable for the investigator to interview each member of the first alarm companies. An account of what each man did upon arrival, what was observed or not noticed, the extent of the fire upon arrival, and the color of the flames and the smoke is always helpful. These observations should be recorded as soon as possible. If necessary, they may be later reduced to written statements. Another point that must be covered in these interviews is whether the observations and statements of the different firefighters agree. Any disagreements or differences in observations must be reconciled.

If the fire is of suspicious origin, were there any signs of evidence outside the building? This would include containers, signs of forcible entry, or someone running or moving away from the fire scene. Discussed in Chapter 4 are the types of facts that the members of the fire department should be communicating to the investigator at this time.

REPORT FORMS

There are numerous types of forms for fire investigation reports, some of great length and detail, others very brief. Each fire investigational unit must develop the type of form that will best serve the needs of its organization.

Taking Notes

Notes are taken for the reason that the investigator cannot remember everything. It is these notes that enable him or her to recall important points of the investigation.

Notes should be as accurate and detailed as possible. Date, time, specific location of the structure, street address, and city or town should be included. If the fire occurred in a multiple occupancy, indicate the specific location and occupant. Where the fire communicated to other sections, occupancies, or structures, record this also. Describe the conditions you observed from outside the structure. Record observations you made upon approach to the scene. Identify the side of the structure where the observations were made to prevent any question as to where you saw what you recorded. Indicate the direction you took when moving around the outside of the structure.

Start the investigation at one specific point in the structure. Depending upon the circumstances of the specific fire, the investigator may start in the basement and work up or begin on the top floor and work down. Each investigation will require a different approach. Upon entering each room or area, record what you observe. Then, as you work the area, record what you see, what you photograph, what you find, what you pick up, what you are looking for. The tape recorder can be of tremendous value to the investigator in this task. Some have compared its use to keeping a diary of the investigation.

Many investigators have found that by the use of both a notebook and a tape recorder, their final reports are much stronger documents. Back at the office, the information from the notebook and the tapes can be transcribed and reconciled. The data thus developed can be used to draft the investigative report. Fire investigation reports should always enumerate the facts and then narrate the findings in an objective manner.

Everything the investigator does or observes should be documented. Do not try to make an entire investigation before recording observations.

NOTE-TAKING RESPONSIBILITIES

The investigator who is working alone will obviously have to take his or her own notes. When the investigator is part of a team, one member of the team will probably be responsible for the note taking. However, if for some reason the team splits up during the investigation, each member will be responsible for notes.

The composition of the investigation team or task force will sometimes determine who will be responsible for note taking. Some teams are made up of members of a fire department, some are strictly a police function. Both kinds have been successful. However, many communities find that a team composed of one member of the fire department and one from the police department has many advantages. In some areas, the fire-department representative does not have the power of arrest, and this can cause a very difficult situation. Having a team representing both the fire and police departments can be a great advantage.

Diagrams

DIAGRAMMING THE SCENE

One of the easier methods of diagramming is by using graph paper. The grid of ten squares to the inch is preferred; since it lends itself to many adaptations. Drawings should be as accurate as possible, but it is not necessary to produce an architectural-style drawing. The main points to be shown are the layout of the area involved and its relation to other areas or rooms. The interior finish and furniture should be indicated. Include on the diagram the area or point of origin and the direction and path of fire travel. How large an undivided area was involved? What was the specific occupancy? If it was an office, where were the desks, tables, machines, and other equipment located? *All measurements must be accurate.*

If a fatality has occurred, the drawing can become even more important. The location of the victim and the position of the body should be indicated. With the finding of a dead body, several questions must be answered: Was the body covered with debris? Was it found on the floor? Was it lying face down or face up? What relation had the dead body's location to the main body of the fire? Was it in an area along the edge of the fire? Was the corpse burned? Was it in a room away from the fire?

In addition to the drawings, either take photographs or have photographs taken to illustrate everything that has been included in the diagrams.

DIAGRAMMING OR DRAWING THE BUILDING

It may be necessary to diagram the entire fire building. Should this occur, it may be possible to obtain copies of blueprints or other architectural drawings from the building and zoning department that has jurisdiction over the area.

Another type of drawing that may be necessary is one of the building and its relation to other buildings in the neighborhood. Some fire departments routinely diagram firefighting operations on all multiple-alarm fires. They include location of engine and truck companies, hose layouts, location of hydrants, and other pertinent information. In addition, such a drawing will show any exposures, if and how the fire communicated, streets, alleys, and similar features.

The investigator or one of the investigation team should do the initial diagrams. Outside help may be obtained, but the maintenance of the integrity of the investigation may be important. For this reason, most investigators prefer to do their own diagramming.

Tools necessary for the initial drawings are simple: pencils, a straightedge, and the graph paper mentioned previously. The final drawing may require also the use of an architectural-style template. This tool helps to properly illustrate door openings, window openings, stairwells, and similar features. Various scales are available: 10, 20, 30, 40, 50, and 60 units to the inch. Fire-prevention drawings and maps have for many years used the scale of one inch to 50 feet, or to 100 feet for larger buildings. Standard symbols for this type of drawing for the fire service are found in both the *NFPA Inspection Manual*[1] and *Fire Prevention and Inspection Practices*.[2] Note that the current fourth edition of the

[1] Charles A. Tuck, Jr., ed., *NFPA Inspection Manual*, 4th ed. (Boston: National Fire Protection Association, 1976), pp. 378–82.

[2] *Fire Prevention and Inspection Practices*, 4th Ed. (Stillwater: International Fire Service Training Association, Oklahoma State University, 1974), pp. 147–49.

Inspection Manual, although it refers to color being used to specify the type of walls, has the diagrams printed in black and white. In the third edition,[3] the same symbols were reproduced in color. Investigators who desire to color-code this type of drawing are therefore referred to the "Standard Plan Symbols" found in the third edition of the *Inspection Manual*, where the various colors are shown.

The investigator must take care to ensure that the measurements made are accurate. Weston and Wells point out:

> The greatest disservice an investigator can render is to record an erroneous measurement on a drawing or in his field notes of a crime, for in that case no matter how many persons subsequently certify the correct measurement, there will always be a question about the initial recording and why it was erroneous.[4]

The representative model · Sometimes, in addition to drawing a diagram of and photographing the fire scene, it may be necessary to construct a representative scale model of the building, its interior and contents. These models are made out of cardboard, posterboard, styrofoam, balsa wood, or any other material that is easy to obtain and to work with. These models are necessary only under special circumstances, such as a fire that causes a large loss of life. If the case requires the construction of such a model, it should probably be prepared by a professional model maker or someone well versed in model making.

Recording the Investigation through Photography

An obvious omission in the list of contents of the fire investigator's kit in Chapter 9 was the camera. This is probably one of the most important tools you will have. If the fire department has a photographic unit, its services may be available to the investigator. However, the investigator should still have his or her own camera.

In the past few years, several very good publications have been released on the subject of fire photography.[5] Most of these discuss "photography at the fire scene" and "photography for the fire investigator." In addition, numerous articles may be found in the various fire-service, law enforcement, and government publications on the subject.

The types of cameras used by investigators vary widely. The authors have used a number of different types from time to time, from the simple fixed-lens camera in the 1930s, to a Kodak Recomar 18 ($2\frac{1}{4} \times 3\frac{1}{4}$ film pack) in the 1940s,

---

[3] Horatio Bond, ed., *NFPA Inspection Manual*, 3rd ed. (Boston: National Fire Protection Association, 1970), inside front and back covers.

[4] Paul B. Weston and Kenneth M. Wells, *Criminal Investigation: Basic Perspectives*, 2nd ed. (Englewood Cliffs, N.J.: Prentice-Hall, 1974), p. 82.

[5] See Bibliography for listing of books on fire photography.

to the 4 × 5 press-type camera after World War II, to the 35mm single-lens reflex and the Polaroid camera system. Today, the investigator has a wide variety of choices and price ranges.

Color photography outperforms the black-and-white process in showing evidence as it is. When shown in court projected on a large screen, color transparencies appear to be superior to prints in fixing the attention of the triers of fact. The quality of life-size reproduction and natural color in this type of photograph goes far beyond what can be achieved by any black-and-white enlarged print. The use of color slides was pioneered by experts in criminalistics, spread to medicolegal experts for autopsy photographs, and is being adopted for on-the-scene photography in many police agencies.[6]

However, under no circumstances should photography be considered a substitute for notes taken at the fire scene. It cannot take the place of accurate measurements and sketches; it simply supplements and reinforces them.

THE EYES OF THE INVESTIGATOR

The camera records what the investigator sees, and sometimes more. As the investigator surveys of the fire scene, the picture taking should begin. Sometimes the circumstances are such that this will not be practical or possible. The first pictures taken will probably be general views of the exterior of the premises, unless the investigator arrives early enough during the fire to record some of the colors of the smoke and flames. Once the origin of the fire has been located, general overall photos of that area must be taken. These should be followed with closeup shots of the point or points of origin.

All suspicious fire patterns at the fire scene should be photographed, and it may be advisable to photograph natural fire patterns there for comparison in court proceedings. Whenever any physical evidence is found, it must be photographed *in place*. Only after being photographed is it removed and preserved, as described in Chapter 9. Then the area where it was found should be rephotographed, to indicate that the evidence was removed. (What constitutes physical evidence was defined and explained in Chapter 9.)

Measurements and measurement markers always have been a problem in crime-scene photography because they intrude upon the photographer's reproduction of the scene as he found it. One acceptable procedure is to take a photograph without any change (as is) and then to take another picture with the measurement marker placed in position.[7]

The condition of all doors and windows pertinent to the investigation should be recorded with photographs. This will indicate whether they were

[6]Weston and Wells, *Criminal Investigation*, pp. 77–78.
[7]*Ibid.*, p. 76.

locked or unlocked prior to the fire. Was any forcible entry made prior to the arrival of the fire department? An example of these circumstances is illustrated in Figures 5-12 and 5-13. If utilities or protective systems, such as sprinkler or alarm systems, have been tampered with, photographs must be taken to show what has been done. Anything that could be any way be considered evidence should be photographed. Photographs of contents and furniture either removed or remaining in the building after the fire may be important.

ADMISSIBILITY OF PHOTOGRAPHIC EVIDENCE

One of the best sources of information and guidance in the admissibility of photographic evidence is an article by the Hon. John A. Decker, a circuit judge in Milwaukee, which was printed in the official publication of the International Association of Arson Investigators, Inc.[8]

Judge Decker pointed out that photographs are qualified for admission into evidence at the discretion of the trial judge. This qualification is usually based upon three requirements: authentication, materiality, and relevancy.

Authentication. Before a photograph may be admitted in evidence, it must be authenticated or verified, that is, its accuracy must be established and its subject identified. Accuracy is a relative matter, for it is impossible, photographically, because of equipment limitation to reproduce a scene which is correct in every minute detail. . . .

. . . Authentication requires that the person, places or things shown in the photograph be identified through the oral testimony of a person, not necessarily the photographer, who can identify the subject matter and state that the reproduction is accurate.[9]

The time when a photograph was taken is important only in connection with any change of the conditions that existed prior to the time it was taken.

Materiality and Relevancy. Photographs are relevant and material if the contents:

(1) relate to the issues being tried,
(2) will aid a witness in illustrating or explaining his testimony, and
(3) will assist the jury in understanding the issues and testimony.[10]

Remember that the fire investigator seeks the true cause of the fire. Photographic evidence, like all other evidence, may exonerate the innocent person as well as strengthening other circumstantial evidence establishing guilt.

[8]Hon. John A. Decker, "The Admissibility of Photographic Evidence," *The Fire and Arson Investigator*, 18, No. 3 (January–March 1968).

[9]*Ibid.*, pp. 40, 41.

[10]*Ibid.*, p. 43.

QUESTIONS

1. Describe the various methods that may be used to record fire investigations.

2. Explain how good fire incident reports and fire inspection reports can be of help to the fire investigator.

3. Describe the three requirements Judge Decker sets forth for qualifying photographs for admission into evidence.

4. Discuss the importance of accurate measurements in connection with taking notes and sketching the area of origin.

5. How may fire incident reports and fire investigation reports reflect upon the public image of the fire department?

6. Describe the step-by-step note-taking procedure you would use at the fire scene.

7. Contrast the use of colored photographic slides with black-and-white photographs in fire investigations.

Techniques of Arson

In order to be able to determine how a fire was set, the investigator must have a working knowledge of how to set a fire. This chapter will be devoted to a discussion of some of the methods used to set fires and some of the means by which one method may be distinguished from another.

One method of obtaining practical experience is to work with local fire departments. Whenever these departments conduct training sessions by burning structures, someone has to set the fire. The investigator may be able to help start the fire and observe its actions and burning patterns under different conditions. By using different fuels, plants, and sets, he or she will soon become familiar with the different patterns developed by various fuels. This knowledge can be of great value at a later date.

Fuels Used in Arson

The fuel most commonly used in arson is gasoline. One of the characteristics of gasoline is that it can explode, a fact that few arsonists realize. The investigator who assists at a training fire should never use straight gasoline. The basic fuel for a training fire should be fuel oil, with just enough gasoline to start it burning. The fuel oil is not nearly as volatile as gasoline; therefore, it does not produce as great a volume of explosive vapors, nor as rapidly.

Gasoline has been mixed with fuel oil on a 50–50 basis and used as an accelerant. At times gasoline may be mixed with soap flakes to make a jelly-like substance. This is not a usable arson tool, but rather an antipersonnel device. It is designed to cause the gasoline to cling to the target and inflict deep burns. Gasoline may also be found in time-delay or remote-ignition systems. When used in this manner, the resultant fire is likely to be accompanied by an explosion. Remote ignition and time delays will be discussed later in this chapter.

Fuel oil is another flammable liquid that can be used as an accelerant. However, it is difficult to ignite and does not spread the fire with the rapidity of the more volatile fuels. When used with gasoline, it complements the gasoline. The fuel oil slows up the explosive properties of gasoline, and the gasoline compensates for the difficult ignition and slow-burning properties of the fuel oil.

Alcohol is another flammable liquid that can be used. It is easily ignited, will burn rapidly, and by itself gives off no detectable smoke. However, unless very large quantities are used, alcohol will not transmit a great deal of heat to the surrounding area. Also alcohol is soluble in water. When it is used as a fuel, the water used to extinguish the fire may carry away all the residual alcohol, making it difficult to recover samples.

Almost any flammable liquid can be used as a fuel. Each has its own set of plus and minus factors. Some will explode; some are very difficult to ignite unless large quantities are used, or some more volatile flammable liquid is used as a primer. Others leave telltale traces or give off detectable smoke and odors. An investigator should purchase samples of the various types of flammable liquids. Sources for obtaining some of these are industrial plants that use these fuels as solvents or in their processes. By setting fires with the different types of fuels, one may observe firsthand the different properties of each. As mentioned previously, this can be done in connection with the training fires conducted by fire departments. Training fires should always be set by two people. One should spread the fuel; the other, ignite it by using a flare or fusee or other remote means.

The investigator should record each of the training fires in which he or she participates. He or she should document them as to the number of fires started by the investigator him- or herself *under controlled conditions*. The reason for this is to assist in qualifying the individual as an "expert witness." The investigator can testify from the witness stand that the purpose of participating in the training fires was to "observe the patterns and characteristics of different fuels under fire conditions." The knowledge obtained from the training fires can then be correlated with what was observed on the scene of the incendiary fire.

The choice of fuels can point toward the skill and ability of the suspect in an arson case. The fuels that are more difficult to obtain, and those of a more complex mixture, are of the type likely to be selected by the professional arsonist. The impulsive, amateurish firesetter who is setting his first fire for profit, or the vandal turned firesetter, will select the more common and available fuels. He or she will show little planning or foresight in the selection and will often use so much of the fuel that traces are left behind for the investigator.

Another easily obtainable fuel that has been used is charcoal igniters—specifically, one that has a wax base. One, called Fire Wax Jelly Starter, is a heavy, waxy liquid that comes in a container similar to a charcoal-lighter fuel can. On the can is the statement, "Firewax is long burning . . . starts all kinds of fires fast." A second type, known as Unita Fire Starter, is a wax-based, semi-solid (putty-like) substance that may be shaped or molded. Materials such as these have been available for several years under different trade names. Fuels of this type have been used as accelerants by applying them to floor joists from below, particularly in unfinished frame structures. The result of these fires is the dropping of entire floors into basements.

Fuel Distribution

The distribution of the fuels and the methods used to place them should receive careful consideration during the investigation. The amateur will usually choose either of two directions of attack. He can splash and spill the accelerant over everything in sight and reach. Or he can attempt to devise a system so intricate and complex that the fire may not even start. The first method, as used by the beginner, leaves burn patterns that should shout "Arson!" to the investigator. This system can be modified by the professional so as to leave little trace. First, the fuel should not be poured on the floor or carpets; the professional will pick furniture, drapes, or flammable trailers suspended off the floor. He will also select a system of ignition and fuel delivery that cannot be found easily. Unless the investigator searches for it, it will remain undetected.

Ignition Systems

In a restaurant fire, the arsonist selected the following ignition system. He removed a heat lamp from the food-warming area at the service counter and put it in an extension cord socket. Then he placed it in a pile of gasoline-soaked paper napkins in the food-service area. The entire area was then doused with gasoline. As the arsonist left, he plugged in the extension cord. The time needed to build up enough heat to ignite the gasoline-soaked napkins was approximately one hour and 20 minutes, as determined by testing this type of system.

After the aforementioned time period, the napkins ignited. However, the time span had allowed the gasoline in the restaurant to vaporize and reach an explosive mixture. The resulting explosion gave the first clue that the fire cause might be arson.

The fire department did an excellent job of extinguishing the fire and preserving evidence. A search of the scene developed the point of origin. From there, the facts began to develop. First, an examination of the six heat-lamp sockets at the food-warming counter revealed that only five of the fixtures still remained in metal screw bases of the heat lamps. This raised the question, Where was the sixth base? A large pile of burned paper napkins was scattered about in the center of the floor. All the other napkins were still in neat packages, some still enclosed in their paper bindings. What was the significance of the one burned pile?

The fire inspector reported that no electrical-code violations had been found at the last inspection. In addition, nowhere in the restaurant had he found any evidence of extension cords being used. Additional examination of the debris revealed that all electrical service in the restaurant was encased in conduit. At the point of origin, the investigators located the copper wires from what appeared to have been an extension cord. These wires led to a wall receptacle, which contained the remains of a male plug such as would be found on an extension cord. At the other end of the wires was a socket, in which was a bulb base. Examination of the base showed it to be the same type as those found in the sockets at the food-warming counter. Where did the extension cord come from, and why was a heat lamp in it? As the answers to these questions were developed, the method used to start the fire was revealed. This gave the investigators direction for their continuing investigation.

Some changes in the arsonist's procedure would have made the investigators' job more difficult. Probably the most important was the selection of fuel. Had a less volatile fuel been used, the explosion—one of the most glaring clues—would have been eliminated. The ignition system was basically sound, but rather old-fashioned.

Time delays of from three seconds to 3,945 minutes are possible by using the circuit chips available at radio shops. They are powered by a nine-volt battery. These circuit chips can be wired to a glow plug or a six-volt bulb with the glass envelope broken off. When they are placed in a pile of fuel-soaked combustibles, a positive ignition will result. Another system can utilize the radio control and receiver from a model airplane. The receiver is placed in the pile of combustibles, together with strike-anywhere matches. By using the control portion, the firesetter can start the fire at the time of his choice and from as far away as half a mile.

The systems of remote ignition or time delay described can be complicated or simple, as desired by the arsonist. On the less complicated side, a plastic bag full of gunpowder and a length of hobby fuze may be used. Whatever the system, each will leave its own characteristic set of remains behind. But only with proper training will the investigator be able to recognize these remains. It is not necessary for the investigator to set up these devices. He or she can put the components at the point of origin and start the training fire. After the fire department has extinguished the fire, he or she can examine what is left. Proper training and a good experience for both the investigator and the firefighters will result in the recovery of a surprising amount of evidence.

CHEMICAL IGNITERS

What are chemical igniters? How difficult are they to obtain? How can they be used? How can the investigator recover useable evidence when igniters have been used? Many chemical compounds, when combined in the proper quantities, will ignite without any outside source of heat. Such ignitions are called hypergolic reactions. The chemicals referred to are not some exotic compounds found only in laboratories. One of the more common is calcium hypochlorite, which was discussed in Chapter 6, under the heading "Oxidizing Agents." Some of the more common fuels that will react with calcium hypochlorite in this manner are brake fluid, charcoal, glycerol, ethyl alcohol, pine oil, turpentine, tobacco, grease, and oil.

Another popular combination is potassium chlorate and sugar. This mixture can be ignited by a fuze or spark, or a heat source, and will provide high heat for a short duration. In addition, when this mixture is used in connection with sulphuric acid, a hypergolic reaction takes place. The mixture of potassium chlorate and sugar leaves behind a deposit of ash. This residue usually contains enough chemicals to provide easily detectable evidence. People who have been in the vicinity of the fire at the time of ignition should be questioned as to what, if anything, they smelled. If someone recalls the odor of marshmallows roasting, it is a strong indication that a sugar-based compound was used.

Whenever potassium chlorate or potassium permanganate is used with acid, a purple stain will be left behind at the point of origin. IT SHOULD NEVER BE TOUCHED BY THE INVESTIGATOR WITH BARE HANDS. When the acid type of device has been used, the stain and residue will contain a large amount of acid. This acid can burn clothing and any unprotected tissue (flesh). In addition, the vapors from this reaction may cause irritation or burns to the eyes and lungs.

The more effective the chemicals are, the more dangerous they are to use. Metallic sodium, potassium, and lithium are very effective when used with water; but they will explode as a result of and during the reaction. All lithium compounds are extremely active and will either burn or explode upon contact with water. Several of them need only the small amounts of water vapor present in air to begin the reaction.

Sodium peroxide, when mixed with an organic fuel—most commonly sugar—becomes a water-reactive hypergolic. It is very effective for starting fires. *WARNING!* This chemical mixture has a side effect that has caused injuries to public-service personnel. The reacting chemicals form a hard, waterproof crust over the unused portion of the mixture. This is a potential source of serious injury to firefighters and investigators. It will remain relatively inert until the crust is broken, usually by someone's stepping on it. When this occurs, the reaction will begin again, with violent results. It has been known to have caused severe burns to the firefighter or investigator and at the same time caused the fire to rekindle. This same mixture may be used to start a secondary and more intense fire. Water is all that is needed to start the reaction.

Another simple but very dangerous mixture is glycerin and potassium permanganate. It not only burns, but will detonate most of the time.

How does an arsonist find out how to use these chemicals? Does he need a background in chemistry to learn these mixtures? No! He needs only to obtain a copy of the NFPA's *Fire Protection Guide on Hazardous Materials*. This book may be found in most fire departments and can be purchased from the NFPA. It was written to disseminate information on the hazardous properties of chemicals to those using the chemicals and to those who might be confronted with emergencies involving them. One section of the *Guide* is the "Manual of Hazardous Chemical Reactions," which cautions about and lists reactions when various chemicals are accidentally mixed or combined in storage or transit. By using this information and by conducting some very risky experiments, the arsonist can become very difficult to identify—either because he is quite good at his trade or because he has been burned crisp in his attempt!

The most difficult type of arson case to work, and yet one that may be likely to yield the most evidence, is the combination of explosives and accelerants. Almost any type of explosive compound can be used with a flammable liquid to increase the destruction and intensify the fire. The result will not be easy to conceal. The evidence and observable patterns will make the *corpus delicti* easy to discover and prove in court.

These devices have been as simple as a can of black powder bound to a bottle of gasoline. These are combined with a fuze or timer to set the black powder off. They can be as complex as 55-gallon drums of gasoline with dynamite charges set to go off in series by radio control. Explosives can also be used to combine chemicals, which in turn will cause a fire.

An arsonist might combine knowledge obtained through his trade or training with just a small amount of imagination. Anyone who has worked in a hospital, a welding shop, or any other place that uses oxygen knows that it must not be permitted to come in contact with any petroleum products. When oxygen does come in contact with a petroleum product, it will ignite and probably detonate. A section of plastic hose may be run from a small container of oxygen into a drum of gasoline. The oxygen-tank valve can be opened by remote control—either electrical or time delay, or by using a long rope. The resultant fire and explosion will destroy most of the evidence, leaving behind a small amount of material. This material will not tell the investigator much unless he knows what he is looking for. The investigator who has not studied his craft will be likely to muddle through an investigation never quite realizing how the act was done.

The "Manual of Hazardous Chemical Reactions" in the NFPA's *Fire Protection Guide* can be of tremendous value to the fire investigator. For instance, under the listing of potassium permanganate, the manual indicates that a violent reaction can be expected when this chemical is mixed with glycerin or glycerol. Combining this information with some imagination, an arsonist might obtain a ten-pound can of potassium permanganate, an eight-ounce bottle of glycerin, a blasting cap, and a length of fuze. He would crimp the cap to the fuze, tape the cap to the bottle, and bury the bottle in the potassium permanganate. If he placed four or five of these in a building with some type of fuel and lit the fuze, the resulting mess would give the unsuspecting investigator one large headache to work through. Once the method is uncovered, however, there can be no doubt as to the incendiary cause of the fire.

The point of these two examples is to direct the fire investigator's thinking toward that of the arsonist. If two or three of these items can be combined to cause a fire or explosion, how can the arsonist combine them with safety to himself? If the investigator was going to combine them, what system could he use? Even more important, what evidence can be expected to be left behind after the reaction and the resulting fire/explosion take place?

In discussing arson techniques, one finds that the range of method is limited only by the imagination of the arsonist. Fortunately, most arsonists are not very original and will resort to spills and a match. But a fire can be and has been started by telephone from a thousand miles away. During the fire investigation, the technique used to set the fire will become apparent. This technique, in turn, will assist in identifying the amount of skill the arsonist can be expected to possess. A particular technique will, in some cases, point to a special skill that the arsonist has. This information can lead the investigator to a suspect. The technique can also point toward the motive for the fire. Was the attack directed at the entire structure? Was it aimed at a specific target within it, such as the records of the company? Was the attack made on some new item of machinery that would be a threat only the competitors of the victim?

Remember, the technique is only one of several pieces of data needed to lead toward the determination of a motive and the establishment of the *corpus delicti*. It is one piece of the jigsaw puzzle that must be properly fitted into the complete picture.

QUESTIONS

1. Discuss the dangers of gasoline when used as an accelerant.

2. When alcohol is used as an accelerant, can it be easily recovered after the fire? Why, or why not?

3. You are setting fire to a structure for a fire training program. You carefully apply the flammable liquid fuel to be used as an accelerant to the plants and trailers you have set up. What is the most practical method *for you* to use to ignite the flammable liquid?

4. What does a purple stain or residue usually indicate? Why should it *never* be touched with the bare hands?

5. Discuss the flammable liquids most commonly used by firesetters.

6. Why should an investigator keep an accurate record of all training fires he or she has set?

7. Describe some of the more common fuels with which calcium hypochlorite will react.

8. What is meant when chemicals are said to be "hypergolic"?

9. How may sodium peroxide mixed with organic fuels be potentially dangerous to the investigator after the fire has been extinguished?

Portable Equipment for the Detection of Fire Accelerants

There are a number of types of portable equipment available to the fire investigator for detecting residues of flammable liquids at the fire scene. Some of these are catalytic combustion detectors, flame ionization detectors, detectors using chemical color tests, gas chromatographs, and infrared spectrophotometers. Each has its limitations, advantages, and disadvantages. Today, new instruments are being developed continuously for this purpose. For the reader who is interested in the capabilities of some of these instruments, an excellent description may be found in the U.S. government publication, *Arson and Arson Investigation: Survey and Assessment*.

In addition, there is the use of olfactory detection—in other words, the sense of smell. Many experienced investigators have trained themselves to recognize the odors of many flammable liquids, the most common being gasoline and fuel oil. There are other problems, however. The nose is not sensitive to some accelerants. There is also something called "olfactory fatigue." This is the fact that the nose can lose its ability to detect some odors after prolonged, intense, or repeated exposures to them. Probably the best known example of this situation is hydrogen sulfide (H_2S). The "rotten-egg" odor of hydrogen sulfide can quickly overcome the ability of the nose to detect it.

As mentioned previously, arsonists have been known to attempt to hide the odor of flammable liquids by using other, particularly strong overpowering odors, such as ammonia, perfumes, and deodorants. Yet the presence of these, when they would not normally be expected, has alerted many an experienced investigator to the possibility that an accelerant was present.

The professional journal for fire investigators recently reported the ThaMac process.[1] By using a electron spectrometer, this process examines the beads formed at the melting point of the copper wires during a fire at the area of

[1]"Did the Short Cause the Fire or Did the Fire Cause the Short," *The Fire and Arson Investigator*, 30, No. 1 (July–September 1979), p. 57.

173

origin. The electron spectrometer bombards the surface of the bead with electrons. Released electrons assist in determining the chemical elements present. The amount of oxygen and the depth to which oxygen has penetrated the bead determines whether the short took place in an oxygen-rich or oxygen-deficient environment. If the short took place in an oxygen-rich environment, it occurred before the fire and was the probable cause of the fire. A fire that is hot enough to melt wires or otherwise cause a short will consume most of the oxygen in the environment around the wires. Consequently, beads from this source will have a relatively small supply of oxygen. This will result in a minimal amount of oxygen being converted into copper oxides, and the amount of penetration of the oxygen into the bead will be very small.

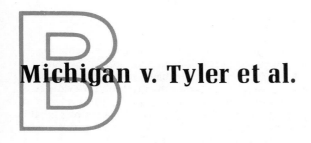

Michigan v. Tyler et al.

CERTIORARI TO THE SUPREME COURT OF MICHIGAN

No. 76–1608. Argued January 10, 1978—Decided May 31, 1978

Shortly before midnight on January 21, 1970, a fire broke out in respondents' furniture store, to which the local fire department responded. When the fire chief arrived at about 2 a. m., as the smoldering embers were being doused, the discovery of plastic containers of flammable liquid was reported to him, and after he had entered the building to examine the containers, he summoned a police detective to investigate possible arson. The detective took several pictures but ceased further investigation because of the smoke and steam. By 4 a. m. the fire had been extinguished and the firefighters departed. The fire chief and detective removed the containers and left. At 8 a. m. the chief and his assistant returned for a cursory examination of the building. About an hour later the assistant and the detective made another examination and removed pieces of evidence. On February 16 a member of the state police arson section took photographs at the store and made an inspection, which was followed by several other visits, at which time additional evidence and information were obtained. Respondents were subsequently charged with conspiracy to burn real property and other offenses. Evidence secured from the building and the testimony of the arson specialist were used at respondents' trial, which resulted in their convictions, notwithstanding their objections that no warrants or consent had been obtained for entries and inspection of the building and seizure of evidentiary items. The State Supreme Court reversed respondents' convictions and remanded the case for a new trial, concluding that "[once] the blaze [has been] extinguished and the firefighters have left the premises, a warrant is required to re-enter and search the premises, unless there is consent or the premises have been abandoned." *Held:*

1. Official entries to investigate the cause of a fire must adhere to the warrant procedures of the Fourth Amendment as made applicable to the States by the Fourteenth Amendment. Since all the entries in this case were "without proper consent" and were not "authorized by a valid search warrant," each one is illegal unless it falls within one of the "certain carefully defined classes of cases" for which warrants are not mandatory. *Camara* v. *Municipal Court*, 387 U. S. 523, 528–529. Pp. 504–509.

175

(a) There is no diminution in a person's reasonable expectation of privacy or in the protection of the Fourth Amendment simply because the official conducting the search is a firefighter rather than a policeman, or because his purpose is to ascertain the cause of a fire rather than to look for evidence of a crime. Searches for administrative purposes, like searches for evidence of crime, are encompassed by the Fourth Amendment. The showing of probable cause necessary to secure a warrant may vary with the object and intrusiveness of the search, but the necessity for the warrant persists. Pp. 505–506.

(b) To secure a warrant to investigate the cause of a fire, an official must show more than the bare fact that a fire occurred. The magistrate's duty is to assure that the proposed search will be reasonable, a determination that requires inquiry into the need for the intrusion, on the one hand, and the threat of disruption to the occupant, on the other. Pp. 506–508.

2. A burning building clearly presents an exigency of sufficient proportions to render a warrantless entry "reasonable," and, once in the building to extinguish a blaze, and for a reasonable time thereafter, firefighters may seize evidence of arson that is in plain view and investigate the causes of the fire. Thus no Fourth and Fourteenth Amendment violations were committed by the firemen's entry to extinguish the blaze at respondents' store, nor by the fire chief's removal of the plastic containers. P. 509.

3. On the facts of this case, moreover, no warrant was necessary for the morning re-entries of the building and seizure of evidence on January 22 after the 4 a. m. departure of the fire chief and other personnel since these were a continuation of the first entry, which was temporarily interrupted by poor visibility. Pp. 510–511.

4. The post-January 22 entries were clearly detached from the initial exigency, and since these entries were made without warrants and without consent, they violated the Fourth and Fourteenth Amendments. Evidence obtained from such entries must be excluded at respondents' retrial. P. 511.

399 Mich. 564, 250 N. W. 2d 467, affirmed.

STEWART, J., delivered the opinion of the Court, in which BURGER, C. J. and POWELL, J., joined; in all but Part IV–A of which WHITE and MARSHALL, J.J., joined; in Parts I, III, and IV of which STEVENS, J., joined and in Parts I, III, and IV–A of which BLACKMUN, J., joined. STEVEN[S], J., filed an opinion concurring in part and concurring in the judgment *post*, p. 512. WHITE, J., filed an opinion concurring in part and dissenting in part, in which MARSHALL, J., joined, *post*, p. 514. REHNQUIST, J., filed a dissenting opinion, *post*, p. 516. BRENNAN, J., took no part in the consideration or decision of the case.

Opinion of the Court

Jeffrey Butler argued the cause *pro hac vice* for petitioner. With him on the brief was *L. Brooks Patterson.*

Jesse R. Bacalis argued the cause and filed a brief for respondents.

MR. JUSTICE STEWART delivered the opinion of the Court.

The respondents, Loren Tyler and Robert Tompkins, were convicted in a Michigan trial court of conspiracy to burn real property in violation of Mich.

Comp. Laws §750.157a (1970).[1] Various pieces of physical evidence and testimony based on personal observation, all obtained through unconsented and warrantless entries by police and fire officials onto the burned premises, were admitted into evidence at the respondents' trial. On appeal, the Michigan Supreme Court reversed the convictions, holding that "the warrantless searches were unconstitutional and that the evidence obtained was therefore inadmissible." 399 Mich. 564, 584, 250 N. W. 2d 467, 477 (1977). We granted certiorari to consider the applicability of the Fourth and Fourteenth Amendments to official entries onto fire-damaged premises. 434 U. S. 814.

I

Shortly before midnight on January 21, 1970, a fire broke out at Tyler's Auction, a furniture store in Oakland County, Mich. The building was leased to respondent Loren Tyler, who conducted the business in association with respondent Robert Tompkins. According to the trial testimony of various witnesses, the fire department responded to the fire and was "just watering down smoldering embers" when Fire Chief See arrived on the scene around 2 a. m. It was Chief See's responsibility "to determine the cause and make out all reports," Chief See was met by Lt. Lawson, who informed him that two plastic containers of flammable liquid had been found in the building. Using portable lights, they entered the gutted store, which was filled with smoke and steam, to examine the containers. Concluding that the fire "could possibly have been an arson," Chief See called Police Detective Webb, who arrived around 3:30 a. m. Detective Webb took several pictures of the containers and of the interior of the store, but finally abandoned his efforts because of the smoke and steam. Chief See briefly "[l]ooked throughout the rest of the building to see if there was any further evidence, to determine what the cause of the fire was." By 4 a. m. the fire had been extinguished and the firefighters departed. See and Webb took the two containers to the fire station, where they were turned over to Webb for safekeeping. There was neither consent nor a warrant for any of these entries into the building, nor for the removal of the containers. The respondents challenged the introduction of these containers at trial, but abandoned their objection in the State Supreme Court. 399 Mich., at 570, 250 N. W. 2d at 470.

Four hours after he had left Tyler's Auction, Chief See returned with Assistant Chief Somerville, whose job was to determine the "origin of all fires that occur within the Township." The fire had been extinguished and the building was empty. After a cursory examination they left, and Somerville returned with Detective Webb around 9 a. m. In Webb's words, they discovered suspicious "burn marks in the carpet, which [Webb] could not see earlier that morning,

[1] In addition, Tyler was convicted of the substantive offenses of burning real property, Mich. Comp. Laws §750.73 (1970), and burning insured property with intent to defraud, Mich. Comp. Laws §750.75 (1970).

because of the heat, steam, and the darkness," They also found "pieces of tape, with burn marks, on the stairway[.]" After leaving the building to obtain tools, they returned and removed pieces of the carpet and sections of the stairs to preserve these bits of evidence suggestive of a fuse trail. Somerville also searched through the rubble "looking for any other signs or evidence that showed how this fire was caused." Again, there was neither consent nor a warrant for these entries and seizures. Both at trial and on appeal, the respondents objected to the introduction of evidence thereby obtained.

On February 16 Sergeant Hoffman of the Michigan State Police Arson Section returned to Tyler's Auction to take photographs.[2] During this visit or during another at about the same time, he checked the circuit breakers, had someone inspect the furnace, and had a television repairman examine the remains of several television sets found in the ashes. He also found a piece of fuse. Over the course of his several visits, Hoffman secured physical evidence and formed opinions that played a substantial role at trial in establishing arson as the cause of the fire and in refuting the respondents' testimony about what furniture had been lost. His entries into the building were without warrants or Tyler's consent, and were for the sole purpose "of making an investigation and seizing evidence." At the trial, respondents' attorney objected to the admission of physical evidence obtained during these visits, and also moved to strike all of Hoffman's testimony "because it was got in an illegal manner."[3]

The Michigan Supreme Court held that with only a few exceptions, any entry onto fire-damaged private property by fire or police officials is subject to the warrant requirements of the Fourth and Fourteenth Amendments. "[Once] the blaze [has been] extinguished and the firefighters have left the premises, a warrant is required to reenter and search the premises, unless there is consent or the premises have been abandoned." 399 Mich., at 583, 250 N. W. 2d, at 477. Applying this principle, the court ruled that the series of warrantless entries that began after the blaze had been extinguished at 4 a. m. on January 22 violated the Fourth and Fourteenth Amendments.[4] It found that the "record

[2]Sergeant Hoffman had entered the premises with other officials at least twice before, on January 26 and 29. No physical evidence was obtained as a result of these warrantless entries.

[3]The State's case was substantially buttressed by the testimony of Oscar Frisch, a former employee of the respondents. He described helping Tyler and Tompkins move valuable items from the store and old furniture into the store a few days before the fire. He also related that the respondents had told him there would be a fire on January 21, and had instructed him to place mattresses on top of other objects so that they would burn better.

[4]Having concluded that warrants should have been secured for the postfire searches, the court explained that different standards of probable cause governed searches to determine the cause of a fire and searches to gather evidence of crime. It then described what standard of probable cause should govern all the searches in this case:

"While it may be no easy task under some circumstances to distinguish as a factual matter between an administrative inspection and a criminal investigation, in the instant case the Court is not faced with that task. Having lawfully discovered the plastic containers of

does not factually support a conclusion that Tyler had abandoned the fire-damaged premises" and accepted the lower court's finding that "'[c]onsent for the numerous searches was never obtained from defendant Tyler.'" *Id.*, at 583, 570-571, 250 N. W. 2d at 476, 470. Accordingly, the court reversed the respondents' convictions and ordered a new trial.

II

The decisions of this Court firmly establish that the Fourth Amendment extends beyond the paradigmatic entry into a private dwelling by a law enforcement officer in search of the fruits or instrumentalities of crime. As this Court stated in *Camara* v. *Municipal Court*, 387 U. S. 523, 528, the "basic purpose of this Amendment . . . is to safeguard the privacy and security of individuals against arbitrary invasions by governmental officials." The officials may be health, fire, or building inspectors. Their purpose may be to locate and abate a suspected public nuisance, or simply to perform a routine periodic inspection. The privacy that is invaded may be sheltered by the walls of a warehouse or other commercial establishment not open to the public. *See* v. *Seattle*, 387 U. S. 541; *Marshall* v. *Barlow's, Inc., ante*, at 311-313. These deviations from the typical police search are thus clearly within the protection of the Fourth Amendment.

The petitioner argues, however, that an entry to investigate the cause of a recent fire is outside that protection because no individual privacy interests are threatened. If the occupant of the premises set the blaze, then, in the words of the petitioner's brief, his "actions show that he has no expectation of privacy" because "he has abandoned those premises within the meaning of the Fourth Amendment." And if the fire had other causes, "the occupants of the premises are treated as victims by police and fire officials." In the petitioner's view, "[t]he likelihood that they will be aggrieved by a possible intrusion into what little remains of their privacy in badly burned premises is negligible."

This argument is not persuasive. For even if the petitioner's contention that arson establishes abandonment be accepted, its second proposition—that innocent fire victims inevitably have no protectible expectations of privacy in whatever remains of their property—is contrary to common experience. People may go on living in their homes or working in their offices after a fire. Even when that is impossible, private effects often remain on the fire-damaged premises. The petitioner may be correct in the view that most innocent fire victims are treated courteously and welcome inspections of their property to ascertain the origin of the blaze, but "even if true, [this contention] is irrelevant to the question whether the . . . inspection is reasonable within the meaning of the Fourth

flammable liquid and other evidence of arson before the fire was extinguished, Fire Chief See focused his attention on assembling proof of arson and began a criminal investigation. At that point there was probable cause for issuance of a criminal investigative search warrant." 399 Mich., at 577, 250 N. W. 2d, at 474 (citations omitted).

Amendment." *Camara, supra*, at 536. Once it is recognized that innocent fire victims retain the protection of the Fourth Amendment, the rest of the petitioner's argument unravels. For it is, of course, impossible to justify a warrantless search on the ground of abandonment by arson when that arson has not yet been proved, and a conviction cannot be used *ex post facto* to validate the introduction of evidence used to secure that same conviction.

Thus, there is no diminution in a person's reasonable expectation of privacy nor in the protection of the Fourth Amendment simply because the official conducting the search wears the uniform of a firefighter rather than a policeman, or because his purpose is to ascertain the cause of a fire rather than to look for evidence of a crime, or because the fire might have been started deliberately. Searches for administrative purposes, like searches for evidence of crime, are encompassed by the Fourth Amendment. And under that Amendment, "one governing principle, justified by history and by current experience, has consistently been followed: except in certain carefully defined classes of cases, a search of private property without proper consent is 'unreasonable' unless it has been authorized by a valid search warrant." *Camara, supra*, at 528–529. The showing of probable cause necessary to secure a warrant may vary with the object and intrusiveness of the search,[5] but the necessity for the warrant persists.

The petitioner argues that no purpose would be served by requiring warrants to investigate the cause of a fire. This argument is grounded on the premise that the only fact that need be shown to justify an investigatory search is that a fire of undetermined origin has occurred on those premises. The petitioner contends that this consideration distinguishes this case from *Camara*, which concerned the necessity for warrants to conduct routine building inspections. Whereas the occupant of premises subjected to an unexpected building inspection may have no way of knowing the purpose or lawfulness of the entry, it is argued that the occupant of burned premises can hardly question the factual basis for fire officials' wanting access to his property. And whereas a magistrate performs the significant function of assuring that an agency's decision to conduct a routine inspection of a particular dwelling conforms with reasonable legislative or administrative standards, he can do little more than rubberstamp an application to search fire-damaged premises for the cause of the blaze. In short, where the justification for the search is as simple and as obvious to everyone as the fact of

---

[5] For administrative searches conducted to enforce local building, health, or fire codes, "'probable cause' to issue a warrant to inspect . . . exist[s] if reasonable legislative or administrative standards for conducting an area inspection are satisfied with respect to a particular dwelling. Such standards, which will vary with the municipal program being enforced, may be based upon the passage of time, the nature of the building (*e.g.*, a multifamily apartment house), or the condition of the entire area, but they will not necessarily depend upon specific knowledge of the condition of the particular dwelling." *Camara*, 387 U. S., at 538; *Marshall* v. *Barlow's, Inc., ante*, at 320–321. See LaFave, Administrative Searches and the Fourth Amendment: The Camara and See Cases, 1967 Sup. Ct. Rev. 1, 18–20.

a recent fire, a magistrate's review would be a time-consuming formality of negligible protection to the occupant.

The petitioner's argument fails primarily because it is built on a faulty premise. To secure a warrant to investigate the cause of a fire, an official must show more than the bare fact that a fire has occurred. The magistrate's duty is to assure that the proposed search will be reasonable, a determination that requires inquiry into the need for the intrusion on the one hand, and the threat of disruption to the occupant on the other. For routine building inspections, a reasonable balance between these competing concerns is usually achieved by broad legislative or administrative guidelines specifying the purpose, frequency, scope, and manner of conducting the inspections. In the context of investigatory fire searches, which are not programmatic but are responsive to individual events, a more particularized inquiry may be necessary. The number of prior entries, the scope of the search, the time of day when it is proposed to be made, the lapse of time since the fire, the continued use of the building, and the owner's efforts to secure it against intruders might all be relevant factors. Even though a fire victim's privacy must normally yield to the vital social objective of ascertaining the cause of the fire, the magistrate can perform the important function of preventing harassment by keeping that invasion to a minimum. See *See* v. *Seattle*, 387 U. S., at 544–545; *United States* v. *Chadwick*, 433 U. S. 1, 9; *Marshall* v. *Barlow's, Inc.*, *ante*, at 323.

In addition, even if fire victims can be deemed aware of the factual justification for investigatory searches, it does not follow that they will also recognize the legal authority for such searches. As the Court stated in *Camara*, "when the inspector demands entry [without a warrant], the occupant has no way of knowing whether enforcement of the municipal code involved requires inspection of his premises, no way of knowing the lawful limits of the inspector's power to search, and no way of knowing whether the inspector himself is acting under proper authorization." 387 U. S., at 532. Thus, a major function of the warrant is to provide the property owner with sufficient information to reassure him of the entry's legality. See *United States* v. *Chadwick*, *supra*, at 9.

In short, the warrant requirement provides significant protection for fire victims in this context, just as it does for property owners faced with routine building inspections. As a general matter, then, official entries to investigate the cause of a fire must adhere to the warrant procedures of the Fourth Amendment. In the words of the Michigan Supreme Court: "Where the cause [of the fire] is undetermined, and the purpose of the investigation is to determine the cause and to prevent such fires from occurring or recurring, a . . . search may be conducted pursuant to a warrant issued in accordance with reasonable legislative or administrative standards or, absent their promulgation, judicially prescribed standards; if evidence of wrongdoing is discovered, it may, of course, be used to establish probable cause for the issuance of a criminal investigative search warrant or in prosecution." But "[i]f the authorities are seeking evidence to be used in a criminal prosecution, the usual standard [of probable cause] will apply." 399 Mich., at 584, 250 N. W. 2d, at 477. Since all the entries in this case were "without proper consent" and were not "authorized by a valid search warrant," each one is illegal unless if falls within one of the "certain carefully defined classes of cases" for which warrants are not mandatory. *Camara*, 387 U. S., at 528–529.

III

Our decisions have recognized that a warrantless entry by criminal law enforcement officials may be legal when there is compelling need for official action and no time to secure a warrant. *Warden* v. *Hayden*, 387 U. S. 294 (warrantless entry of house by police in hot pursuit of armed robber); *Ker* v. *California*, 374 U. S. 23 (warrantless and unannounced entry of dwelling by police to prevent imminent destruction of evidence). Similarly, in the regulatory field, our cases have recognized the importance of "prompt inspections, even without a warrant, . . . in emergency situations." *Camara, supra*, at 539, citing *North American Cold Storage Co.* v. *Chicago*, 211 U. S. 306 (seizure of unwholesome food); *Jacobson* v. *Massachusetts*, 197 U. S. 11 (compulsory smallpox vaccination); *Compagnie Francaise* v. *Board of Health*, 186 U. S. 380 (health quarantine).

A burning building clearly presents an exigency of sufficient proportions to render a warrantless entry "reasonable." Indeed, it would defy reason to suppose that firemen must secure a warrant or consent before entering a burning structure to put out the blaze. And once in a building for this purpose, firefighters may seize evidence of arson that is in plain view. *Coolidge* v. *New Hampshire*, 403 U. S. 443, 465–466. Thus, the Fourth and Fourteenth Amendments were not violated by the entry of the firemen to extinguish the fire at Tyler's Auction, nor by Chief See's removal of the two plastic containers of flammable liquid found on the floor of one of the showrooms.

Although the Michigan Supreme Court appears to have accepted this principle, its opinion may be read as holding that the exigency justifying a warrantless entry to fight a fire ends, and the need to get a warrant begins, with the dousing of the last flame. 399 Mich., at 579, 250 N. W. 2d, at 475. We think this view of the firefighting function is unrealistically narrow, however. Fire officials are charged not only with extinguishing fires, but with finding their causes. Prompt determination of the fire's origin may be necessary to prevent its recurrence, as through the detection of continuing dangers such as faulty wiring or a defective furnace. Immediate investigation may also be necessary to preserve evidence from intentional or accidental destruction. And, of course, the sooner the officials complete their duties, the less will be their subsequent interference with the privacy and the recovery efforts of the victims. For these reasons, officials need no warrant to remain in a building for a reasonable time to investigate the cause of a blaze after it has been extinguished.[6] And if the warrantless entry to put out the fire and determine its cause is constitutional, the warrantless seizure of evidence while inspecting the premises for these purposes also is constitutional.

IV

A

The respondents argue, however, that the Michigan Supreme Court was correct in holding that the departure by the fire officials from Tyler's Auction at 4 a. m. ended any license they might have had to conduct a warrantless search. Hence, they say that even if the firemen might have been entitled to remain in the building without a warrant to investigate the cause of the fire, their departure and re-entry four hours later that morning required a warrant.

On the facts of this case, we do not believe that a warrant was necessary for the early morning re-entries on January 22. As the fire was being extinguished, Chief See and his assistants began their investigation, but visibility was severely hindered by darkness, steam, and smoke. Thus they departed at 4 a. m. and returned shortly after daylight to continue their investigation. Little purpose would have been served by their remaining in the building, except to remove any doubt about the legality of the warrantless search and seizure later that same morning. Under these circumstances, we find that the morning entries were no more than an actual continuation of the first, and the lack of a warrant thus did not invalidate the resulting seizure of evidence.

B

The entries occurring after January 22, however, were clearly detached from the initial exigency and warrantless entry. Since all of these searches were conducted without valid warrants and without consent, they were invalid under the Fourth and Fourteenth Amendments, and any evidence obtained as a result of those entries must, therefore, be excluded at the respondents' retrial.

---

[6]The circumstances of particular fires and the role of firemen and investigating officials will vary widely. A fire in a single-family dwelling that clearly is extinguished at some identifiable time presents fewer complexities than those likely to attend a fire that spreads through a large apartment complex or that engulfs numerous buildings. In the latter situations, it may be necessary for officials—pursuing their duty both to extinguish the fire and to ascertain its origin—to remain on the scene for an extended period of time repeatedly entering or re-entering the building or buildings, or portions thereof. In determining what constitutes a "reasonable time to investigate," appropriate recognition must be given to the exigencies that confront officials serving under these conditions, as well as to individuals' reasonable expectations of privacy.

V

In summation, we hold that an entry to fight a fire requires no warrant, and that once in the building, officials may remain there for a reasonable time to investigate the cause of the blaze. Thereafter, additional entries to investigate the cause of the fire must be made pursuant to the warrant procedures governing administrative searches. See *Camara*, 387 U. S., at 534-539; *See* v. *Seattle*, 387 U. S., at 544-545; *Marshall* v. *Barlow's, Inc.*, *ante*, at 320-321. Evidence of arson discovered in the course of such investigations is admissible at trial, but if the investigating officials find probable cause to believe that arson has occurred and require further access to gather evidence for ·a possible prosecution, they may obtain a warrant only upon a traditional showing of probable cause applicable to searches for evidence of crime. *United States* v. *Ventresca*, 380 U. S. 102.

These principles require that we affirm the judgment of the Michigan Supreme Court ordering a new trial.[7]

Affirmed.

Opinion of STEVENS, J.

MR. JUSTICE BLACKMUN joins the judgment of the Court and Parts I, III, and IV-A of its opinion.

MR. JUSTICE BRENNAN took no part in the consideration or decision of this case.

MR. JUSTICE STEVENS, concurring in part and concurring in the judgment.

Because Part II of the Court's opinion in this case, like the opinion in *Camara* v. *Municipal Court*, 387 U. S. 523, seems to assume that an official search must either be conducted pursuant to a warrant or not take place at all, I cannot join its reasoning.

In particular, I cannot agree with the Court's suggestion that, if no showing of probable cause could be made, "the warrant procedures governing administrative searches," *ante*, at 511, would have complied with the Fourth Amendment.

---

[7]The petitioner alleges that respondent Tompkins lacks standing to object to the unconstitutional searches and seizures. The Michigan Supreme Court refused to consider the State's argument, however, because the prosecutor failed to raise the issue in the trial court or in the Michigan Court of Appeals. 399 Mich., at 571, 250 N. W. 2d, at 470-471. We read the state court's opinion to mean that in the absence of a timely objection by the State, a defendant will be presumed to have standing. Failure to present a federal question in conformance with the state procedure constitutes an adequate and independent ground of decision barring review in this Court, so long as the State has a legitimate interest in enforcing its procedural rule. *Henry* v. *Mississippi*, 379 U. S. 443, 447. See *Safeway Stores* v. *Oklahoma Grocers*, 360 U. S. 344 342 n. 7; *Cardinale* v. *Louisiana* 394 U. S. 437, 438. The petitioner does not claim that Michigan's procedural rule serves no legitimate purpose. Accordingly, we do not entertain the petitioner's standing claim which the state court refused to consider because of procedural default.

In my opinion, an "administrative search warrant" does not satisfy the requirements of the Warrant Clause.[1] See *Marshall* v. *Barlow's, Inc., ante,* p. 325 (STEVENS, J., dissenting). Nor does such a warrant make an otherwise unreasonable search reasonable.

A warrant provides authority for an unannounced, immediate entry and search. No notice is given when an application for a warrant is made and no notice precedes its execution; when issued, it authorizes entry by force.[2] In my view, when there is no probable cause to believe a crime has been committed and when there is no special enforcement need to justify an unannounced entry,[3] the Fourth Amendment neither requires nor sanctions an abrupt and peremptory confrontation between sovereign and citizen.[4] In such a case, to comply with the constitutional requirement of reasonableness, I believe the sovereign must provide fair notice of an inspection.[5]

The Fourth Amendment interests involved in this case could have been protected in either of two ways—by a warrant, if probable cause existed; or by fair notice, if neither probable cause nor a special law enforcement need existed. Since the entry on February 16 was not authorized by a warrant and not preceded by advance notice, I concur in the Court's judgment and in Parts I, III, and IV of its opinion.

---

[1] The Warrant Clause of the Fourth Amendment provides that "no Warrants shall issue, but upon probable cause, supported by Oath or affirmation, and particularly describing the place to be searched, and the persons or things to be seized."

[2] See *Wyman* v. *James,* 400 U. S. 309, 323–324. As the Court observed in *Wyman,* a warrant is not simply a device providing procedural protections for the citizen; it also grants the government increased authority to invade the citizen's privacy. See *Miller* v. *United States,* 357 U. S. 301, 307–308.

[3] In this case, there obviously was a special enforcement need justifying the initial entry to extinguish the fire, and I agree that the search on the morning after the fire was a continuation of that entirely legal entry. A special enforcement need can, of course, be established on more than a case-by-case basis, especially if there is a relevant legislative determination of need. See *Marshall* v. *Barlow's, Inc., ante,* p. 325 (STEVENS, J., dissenting).

[4] The Fourth Amendment ensures "[t]he right of the people to be *secure* in their persons, houses, papers, and effects, against unreasonable searches and seizures." (Emphasis added.) Surely this broad protection encompasses the expectation that the government cannot demand immediate entry when it has neither probable cause to suspect illegality nor any other pressing enforcement concern. Yet under the rationale in Part II of the Court's opinion, the less reason an officer has to suspect illegality, the less justification he need give the magistrate in order to conduct an unannounced search. Under this rationale, the police will have no incentive—indeed they have a disincentive—to establish probable cause before obtaining authority to conduct an unannounced search.

[5] See LaFave, Administrative Searches and the Fourth Amendment: The Camara and See Cases, 1967 Sup. Ct. Rev. 1. The requirement of giving notice before conducting a routine administrative search is hardly unprecedented. It closely parallels existing procedures for administrative subpoenas, see, *e.g.,* 15 U. S. C. § 1312 (1976 ed.), and is, as Professor LaFave points out, embodied in English law and practice. See LaFave, *supra,* at 31–32.

Opinion of WHITE, J.

MR. JUSTICE WHITE, with whom MR. JUSTICE MARSHALL joins, concurring in part and dissenting in part.

I join in all but Part IV–A of the opinion, from which I dissent. I agree with the Court that:

> "[A]n entry to fight a fire requires no warrant, and that once in the building, officials may remain there for a reasonable time to investigate the cause of the blaze. Thereafter, additional entries to investigate the cause of the fire must be made pursuant to the warrant procedures governing administrative searches." *Ante,* at 511.

The Michigan Supreme Court found that the warrantless searches, at 8 and 9 a. m. were not, in fact, continuations of the earlier entry under exigent circumstances* and therefore ruled inadmissible all evidence derived from those searches. The Court offers no sound basis for overturning this conclusion of the state court that the subsequent re-entries were distinct from the original entry. Even if, under the Court's "reasonable time" criterion, the firemen might have stayed in the building for an additional four hours—a proposition which is by no means clear—the fact remains that the firemen did not choose to remain and continue their search, but instead locked the door and departed from the premises entirely. The fact that the firemen were willing to leave demonstrates that the exigent circumstances justifying their original warrantless entry were no longer present. The situation is thus analogous to that in *G. M. Leasing Corp.* v. *United States,* 429 U. S. 338, 358–359 (1977):

> The agents' own action . . . in their delay for two days following their first entry, and for more than one day following the observation of materials being moved from the office, before they made the entry during which they seized the records, is sufficient to support the District Court's implicit finding that there were no exigent circumstances. . . .

To hold that some subsequent re-entries are "continuations" of earlier ones will not aid firemen, but confuse them, for it will be difficult to predict in advance how a court might view a re-entry. In the end, valuable evidence may be excluded for failure to seek a warrant that might have easily been obtained.

Those investigating fires and their causes deserve a clear demarcation of the

---

*The Michigan Supreme Court recognized that "[i]f there are exigent circumstances, such as reason to believe that the destruction of evidence is imminent or that a further entry of the premises is necessary to prevent the recurrence of the fire, no warrant is required and evidence discovered is admissible." 399 Mich. 564, 578, 250 N. W. 2d 467, 474 (1977). It found, however, that "[i]n the instant case there were no exigent circumstances justifying the searches made hours, days or weeks after the fire was extinguished," *Id.,* at 579, 250 N. W. 2d, at 475.

constitutional limits of their authority. Today's opinion recognizes the need for speed and focuses attention on fighting an ongoing blaze. The firetruck need not stop at the courthouse in rushing to the flames. But once the fire has been extinguished and the firemen have left the premises, the emergency is over. Further intrusion on private property can and should be accompanied by a warrant indicating the authority under which the firemen presume to enter and search.

There is another reason for holding that re-entry after the initial departure required a proper warrant. The state courts found that at the time of the first re-entry a criminal investigation was under way and that the purpose of the officers in re-entering was to gather evidence of crime. Unless we are to ignore these findings, a warrant was necessary. *Camara* v. *Municipal Court*, 387 U. S. 523 (1967), and *See* v. *Seattle*, 387 U. S. 541 (1967), did not differ with *Frank* v. *Maryland*, 359 U. S. 360 (1959), that searches for criminal evidence are of special significance under the Fourth Amendment.

REHNQUIST, J., dissenting

MR. JUSTICE REHNQUIST, dissenting.

I agree with my Brother STEVENS, for the reasons expressed in his dissenting opinion in *Marshall* v. *Barlow's, Inc.*, *ante*, at 328 (STEVENS, J., dissenting), that the "Warrant Clause has no application to routine, regulatory inspections of commercial premises." Since in my opinion the searches involved in this case fall within that category, I think the only appropriate inquiry is whether they were reasonable. The Court does not dispute that the entries which occured at the time of the fire and the next morning were entirely justified, and I see nothing to indicate that the subsequent searches were not also eminently reasonable in light of all the circumstances.

In evaluating the reasonableness of the later searches, their most obvious feature is that they occurred after a fire which had done substantial damage to the premises, including the destruction of most of the interior. Thereafter the premises were not being used and very likely could not have been used for business purposes, at least until substantial repairs had taken place. Indeed, there is no indication in the record that after the fire Tyler ever made any attempt to secure the premises. As a result, the fire department was forced to lock up the building to prevent curious bystanders from entering and suffering injury. And as far as the record reveals, Tyler never objected to this procedure or attempted to reclaim the premises for himself.

Thus, regardless of whether the premises were technically "abandoned" within the meaning of the Fourth Amendment, cf. *Abel* v. *United States*, 362 U. S. 217, 241 (1960); *Hester* v. *United States*, 265 U. S. 57 (1924), it is clear to me that no purpose would have been served by giving Tyler notice of the intended search or by requiring that the search take place during the hours which in other situations might be considered the only "reasonable" hours to conduct a regulatory search. In fact, as I read the record, it appears that Tyler not only had notice that the investigators were occasionally entering the premises for the purpose of determining the cause of the fire, but he never voiced the slightest objection to these searches and actually accompanied the investigators on at least one occasion. App. 54–57. In fact, while accompanying the investigators during one of these searches, Tyler himself suggested that the fire very well may have been caused by arson. *Id*., at 56. This observation, coupled with all the other circumstances, including Tyler's knowledge of, and apparent acquiescence in, the searches, would have been taken by any sensible person as an indication that Tyler thought the searches ought to continue until the culprit was discovered; at the very least they indicated that he had no objection to these searches. Thus, regardless of what sources may serve to inform one's sense of what is reasonable, in the circumstances of this case I see nothing to indicate that these searches were in any way unreasonable for purposes of the Fourth Amendment.

Since the later searches were just as reasonable as the search the morning immediately after the fire in light of all these circumstances, the admission of evidence derived therefrom did not, in my opinion, violate respondents' Fourth and Fourteenth Amendment rights. I would accordingly reverse the judgment of the Supreme Court of Michigan which held to the contrary.

Bibliography

Bond, Horatio, ed., *NFPA Inspection Manual*, 3rd ed. Boston: National Fire Protection Association, 1970.

Boudreau, John F., Quon Y. Kwan, William E. Faragher, and **Genevieve C. Denault,** *Arson and Arson Investigation: Survey and Assessment.* Washington, D.C.: U.S. Department of Justice, National Institute of Law Enforcement and Criminal Justice, Law Enforcement Assistance Administration, 1977.

Conway, C. W., "Incendiary Fires in Industrial Occupancies," *Fire Journal*, 70, No. 2 (March 1976), 28–33.

Curtis, Arthur F., *A Treatise on the Law of Arson.* Buffalo, N.Y.: Dennis & Co., Inc., 1936.

Edland, John F., M.D., "Fire Victims," in *Forensic Pathology – A Handbook for Pathologists*, eds. Russel S. Fisher, M.D., and Charles S. Petty, M.D. Washington, D.C.: Department of Justice, National Institute of Law Enforcement and Criminal Justice, 1977.

Federal Bureau of Investigation, *Handbook of Forensic Science.* Washington, D.C.: U.S. Government Printing Office, 1975.

Firemanship Training, *Test Fire, Des Moines, Iowa, November 14, 1959.* Ames: Engineering Extension, Iowa State University, n.d.

Fisher, Russell S., M.D., and **Charles S. Petty, M.D.,** eds., *Forensic Pathology – A Handbook for Pathologists.* Washington, D.C.: U.S. Department of Justice, National Institute of Law Enforcement and Criminal Justice, Law Enforcement Assistance Administration, 1977.

International Association of Arson Investigators, *Selected Articles for Fire and Arson Investigators.* Marlboro, Mass.: International Association of Arson Investigators, Inc., 1975.

International Fire Service Training Association, *Fire Prevention and Inspection Practices*, Stillwater: Fire Protection Publications, Oklahoma State University, 1974.

——, *Photography for the Fire Service.* Stillwater: Fire Protection Publications, Oklahoma State University, 1977.

Johnson, Allen J., "Fingerprints of Fire – Investigating Explosive Fire Causes,"

Fuels, Fuels Utilization & Evaluation, Letters & Service. Lansdowne, Pa.: By the author, 1958.

Levin, Bernard, "Psychological Characteristics of Firesetters," *Fire Journal,* 70, No. 2 (March 1976), 36–41.

Lewis, Nolan D. C., and **Helen Yarnell,** *Pathological Firesetting (Pyromania).* New York: Coolidge Foundation, 1951.

Lewis, Peter W., *Criminal Procedures: The Supreme Court's View–Cases,* pp. 280–89. St. Paul: West Publishing Company, 1979.

Lyons, Paul Robert, *Techniques of Fire Photography.* Boston: NFPA Fire Protection Association, 1978.

McKinnon, Gordon P., ed., *Fire Protection Handbook,* 14th ed. Boston: National Fire Protection Association, 1976.

Meidl, James H., *Explosive and Toxic Hazardous Materials.* Encino: Glencoe Press, 1970.

———, *Flammable Hazardous Materials,* 2nd ed. Encino: Glencoe Publishing Co., 1978.

Meyer, Eugene, *Chemistry of Hazardous Materials.* Englewood Cliffs: Prentice-Hall, 1977.

National Automobile Theft Bureau, *Manual for the Investigation of Automobile Fires.* Jericho: National Automobile Theft Bureau, 1977.

Oakes, Kent A., "The Role of the Crime Laboratory in the Prosecution of an Arson Case," paper presented at the 12th Annual Wisconsin Arson Seminar, Madison, Wis., June 8, 1977.

Rethoret, Harry, *Fire Investigations.* Montreal: Recording and Statistical Corporation, Ltd., 1945.

Schaffer, E. L., *Charring Rate of Selected Woods–Transverse to Grain,* Research Paper FPL69. Madison: U.S. Forest Products Laboratory, 1967.

———, *Review of Information Related to the Charring Rate of Wood,* U.S. Forest Service Research Note FPL-0145. Madison: U.S. Forest Products Laboratory, 1966.

Schmidt, Wayne W., ed., *Fire and Police Personnel Reporter.* South San Francisco: Public Safety Personnel Research Institute, Inc. An excellent monthly publication that reviews recent court decisions, significant arbitration awards, administrative appeals, selected pending litigation, and other matters concerning fire and police personnel.

Seattle Fire Department, *Seattle Arson Task Force.* Seattle: Seattle Fire Department, n.d.

Simson, Laurence R., Jr., M.D., "Fatal Fire Investigations," paper presented at the 13th Annual Wisconsin Arson Seminar, LaCrosse, Wis., June 7, 1978.

Spiegelman, Arthur, "Autopsy of an Explosion," *Fire Engineering,* 120, No. 10 (October 1967), 52–53.

Spitz, W. V., M.D., and **R. S. Fisher, M.D.,** *Medicolegal Investigations of Death.* Springfield: Charles C Thomas, 1973.

Stekel, Wilhelm, "Pyromania," in *Peculiarities of Behavior,* 2, XI, 124–81, trans. James S. Van Teslaar. New York: Liveright Publishing Corporation, 1924; Library Edition, 1943.

Stuerwald, John E., ed., *The Fire and Arson Investigator.* Marlboro: International Association of Arson Investigators, Inc. Quarterly publication of the International Association of Arson Investigators. Contains excellent material and information for the fire investigator.

Suchy, John T., ed., *Arson: America's Malignant Crime*, final report prepared for National Fire Prevention and Control Administration, U.S. Department of Commerce. Washington, D.C.: U.S. Government Printing Office, 1976.

Sutton, Dr. Beverly, "Pyromania and Psychopathic Firesetters," *The Fire and Arson Investigator*, 25, No. 2 (October–December 1974), 23–36.

Tuck, Charles A., Jr., ed., *NFPA Inspection Manual*, 4th ed. Boston: National Fire Protection Association, 1976.

U.S. Forest Products Laboratory, *Ignition and Charring Temperatures of Wood*, Report No. 1464, rev. Madison: U.S. Forest Products Laboratory, 1958.

Weston, Paul B., and **Kenneth M. Wells**, *Criminal Investigation: Basic Perspectives*, 3rd ed. Englewood Cliffs: Prentice-Hall, 1980.

Index

Index